Analog Circuits and Signal Processing

The Analog Circuits and Signal Processing book series, formerly known as the Kluwer International Series in Engineering and Computer Science, is a high level academic and professional series publishing research on the design and applications of analog integrated circuits and signal processing circuits and systems. Typically per year we publish between 5-15 research monographs, professional books, handbooks, edited volumes and textbooks with worldwide distribution to engineers, researchers, educators, and libraries.

The book series promotes and expedites the dissemination of new research results and tutorial views in the analog field. There is an exciting and large volume of research activity in the field worldwide. Researchers are striving to bridge the gap between classical analog work and recent advances in very large scale integration (VLSI) technologies with improved analog capabilities. Analog VLSI has been recognized as a major technology for future information processing. Analog work is showing signs of dramatic changes with emphasis on interdisciplinary research efforts combining device/circuit/technology issues. Consequently, new design concepts, strategies and design tools are being unveiled.

Topics of interest include:

Analog Interface Circuits and Systems;

Data converters;

Active-RC, switched-capacitor and continuous-time integrated filters;

Mixed analog/digital VLSI;

Simulation and modeling, mixed-mode simulation;

Analog nonlinear and computational circuits and signal processing;

Analog Artificial Neural Networks/Artificial Intelligence;

Current-mode Signal Processing; Computer-Aided Design (CAD) tools;

Analog Design in emerging technologies (Scalable CMOS, BiCMOS, GaAs, heterojunction and floating gate technologies, etc.);

Analog Design for Test;

Integrated sensors and actuators; Analog Design Automation/Knowledge-based Systems; Analog VLSI cell libraries; Analog product development; RF Front ends, Wireless communications and Microwave Circuits;

Analog behavioral modeling, Analog HDL.

More information about this series at http://www.springer.com/series/7381

Gürkan Yılmaz · Catherine Dehollain

Wireless Power Transfer and Data Communication for Neural Implants

Case Study: Epilepsy Monitoring

Gürkan Yılmaz
EPFL RFIC Research Group
Lausanne
Switzerland

Catherine Dehollain
EPFL RFIC Research Group
Lausanne
Switzerland

ISSN 1872-082X ISSN 2197-1854 (electronic)
Analog Circuits and Signal Processing
ISBN 978-3-319-84137-3 ISBN 978-3-319-49337-4 (eBook)
DOI 10.1007/978-3-319-49337-4

Printed on acid-free paper

This Springer imprint is published by Springer Nature
The registered company is Springer International Publishing AG
The registered company address is: Gewerbestrasse 11, 6330 Cham, Switzerland

To my family
Kiraz, Hilmi and Serkan

and

To my love Eva-Liisa

—Gürkan Yılmaz

Preface

Recording neural activities plays an important role in numerous applications ranging from brain mapping to implementation of brain–machine interfaces (BMIs) to recover lost functions or to understand the mechanisms behind the neurological disorders such as essential tremor, Parkinson's disease, and epilepsy. It also constitutes the first step of a closed-loop therapy system which employs a stimulator and a decision mechanism additionally. Such systems are envisaged to record neural anomalies and then stimulate corresponding tissues to cease such activities.

Methods for recording the neural signals have evolved to its current state since decades, and the evolution still goes on. This book focuses on how to eliminate all the wired connections for new-generation neural recording systems: implantable wireless neural recording systems with a case study on in vivo epilepsy monitoring. The scope of this book can be defined as wireless power transfer, wireless data communication, biocompatible packaging, and compulsory experiments on the way to human trials.

First of all, wireless power transmission is realized using 4-coil resonant inductive link topology which exploits the magnetic coupling phenomena. In addition to the power transfer, a reliable DC power supply is generated in the implant by means of a half-wave active rectifier and a low drop-out voltage regulator. The operation frequency, 8.5 MHz, has been optimized by taking tissue absorption and bandwidth limitations for data communication into account.

Secondly, wireless data communication solutions have been investigated, and two different solutions have been implemented for different application scenarios: First solution is to use load modulation scheme, which actually relies on varying the load according to the incoming neural data. However, there is a trade-off between data rate and power transfer efficiency for this solution, which in return leads us to implement the second solution, dedicated transmitter at a higher frequency. Consequently, a transmitter which can work at MICS (402–205 MHz), ISM (433 MHz), and several MedRadio bands has been implemented to transmit neural data to an external base station which includes a discrete receiver.

Following the integration of all electronic circuits which have been fabricated using UMC 180-nm MM/RF technology, the implant has been packaged using biocompatible polymers (PDMS, medical grade epoxy, and Parylene-C). Packaging provides the bidirectional diffusion barrier feature which enables in vitro and in vivo experiments to be conducted.

Finally, three levels of experimentation have been conducted to validate the operation of the system: in air for electrical characterization, in a tissue-mimicking solution for in vitro characterization, and in a mouse brain for in vivo characterization.

Lausanne, Switzerland
September 2016

Gürkan Yılmaz
Catherine Dehollain

Contents

Acronyms

AC	Alternating current
ADC	Analog-to-digital converter
ASK	Amplitude-shift keying
BER	Bit error rate
BW	Bandwidth
CMOS	Complementary metal–oxide–semiconductor
DAC	Digital-to-analog converter
DC	Direct current
DR	Data rate
E	Electrical field strength
EA	Error amplifier
ECG	Electrocardiography
ECoG	Electrocorticography
EEG	Electroencephalography
EIRP	Effective isotropic radiated power
ESA	Electrically small antenna
FDA	Food and Drug Administration
FoM	Figure of merit
FPGA	Field-programmable gate array
FSK	Frequency-shift keying
H	Magnetic field strength
IC	Integrated circuit
iEEG	Intracranial electroencephalography
ISM	Industrial, Scientific and Medical
kbps	Kilobit per second
LDO	Low drop-out
LSK	Load-shift keying
Mbps	Megabit per second
MEA	Microelectrode array
MI	Modulation index

MIM	Metal–insulator–metal
MOSFET	Metal–oxide–semiconductor field-effect transistor
MUA	Multi-unit activity
OOK	On–off keying
OPAMP	Operational amplifier
OTA	Operational transconductance amplifier
PA	Power amplifier
PCB	Printed circuit board
PSR	Power supply rejection
PSRR	Power supply rejection ratio
PTE	Power transfer efficiency
PZT	Lead zirconate titanate (piezoelectric ceramic material)
RF	Radio frequency
RFID	Radio frequency identification
RMS	Root mean square
RX	Receiver
SAR	Specific absorption rate
SNR	Signal-to-noise ratio
SoC	System on chip
SRF	Self-resonant frequency
TX	Transmitter
VCO	Voltage-controlled oscillator
WHO	World Health Organization
WPT	Wireless power transfer

Chapter 1
Introduction

Abstract "How does the brain work?" is one of the Yilmaz questions that has been asked the most throughout the history of medical sciences. The question is addressed from different perspectives by all branches of science, in particular by the life sciences. Although all seek a different answer, there is something common which triggers all these researches: *observation*. With curiosity, thats what makes the beginning of a scientific study. From electrical engineering perspective, observation regarding this question is performed via recording the electrical signals generated by the neurons and interpreting those results. Recording neural activities plays an important role in numerous applications ranging from brain mapping to implementation of brain-machine interfaces (BMI) to recover lost functions or to understand the mechanisms behind the neurological disorders such as essential tremor, Parkinsons disease [1] and epilepsy [2, 3]. It also constitutes the first step of a closed-loop therapy system which additionally employs a stimulator and a decision mechanism. Such systems are envisaged to record neural anomalies and then stimulate corresponding tissues to cease such activities. Methods for recording the neural signals have evolved to its current state since decades, and the evolution still goes on. This chapter introduces the fundamentals of the new generation neural recording systems: implantable wireless neural recording systems with a case study on in-vivo epilepsy monitoring. Throughout this chapter, firstly, we have defined the problem and then, the motivation to solve this problem, introducing the systems anticipated benefits. Next, challenges to be encountered while realizing such a system have been explained and our approach has been briefly introduced.

1.1 Problem Definition

Current clinical practice in recording electrical activities of the brain is dominated by electroencephalography (EEG) which is a non-invasive procedure performed alongthe scalp. Another type of EEG, intracranial EEG (iEEG; *aka* electrocorticog-

G. Yılmaz and C. Dehollain, *Wireless Power Transfer and Data Communication for Neural Implants*, Analog Circuits and Signal Processing,
DOI 10.1007/978-3-319-49337-4_1

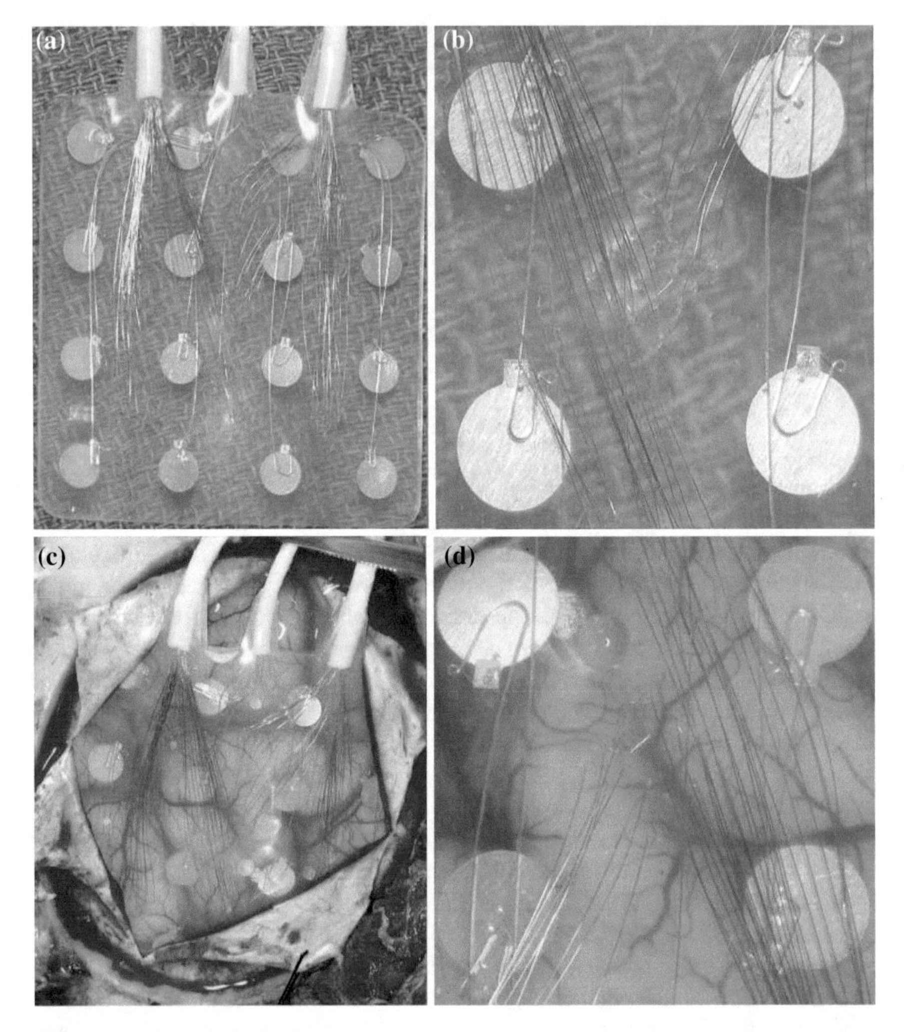

Fig. 1.1 Depictions of the hybrid grid for intracranial monitoring: **a** low magnification view of hybrid 4 × 4 grid prior to implantation, **b** high magnification view of the grid subunit, **c** low magnification view of implanted grid in a patient, and **d** high magnification view of grid subunit over human cortex [4]

raphy (ECoG)) is an invasive procedure which is performed by placing an electrode matrix (or array) onto the cortex following the craniotomy as presented in Fig. 1.1 [4]. Intracranial EEG is, for instance, employed for epileptic focus localization prior to the resective surgery [5] which is performed to treat certain types of epilepsy.

Currently, iEEG is performed by using electrode arrays composed of millimeter-sized passive electrodes [6] as already shown in Fig. 1.1. Neural signals acquired via these electrodes are transferred to an external recording device via transcutaneous wires. However, the wired connection through the skull during the monitoring period increases the risk of the cerebrospinal fluid (CSF) leakage, or worse, infection of the CSF. So far, all the implantation procedures are achieved using transcutaneous wired devices, associated with potential serious complications in up to 25% of cases, such as intracranial infections or CSF leakage [7]. Finally, patients are permanently connected to a recording station through multiple connecting cables, leading to decreased comfort and autonomy. This situation reduces the patient mobility and affects psychological state of the patient and more importantly limits the monitoring period. Sealing the craniotomy area up is anticipated to eliminate or at least minimize these risks. Consequently, it can be foreseen that employing a wireless data transmission instead of wired connection will overcome the aforementioned issues.

As the target application of this book, next section gives an introduction to epilepsy and its surgical treatment to have a better understanding of the problem and a more precise definition of the problems encountered due to the current clinical practice.

Epilepsy and Surgical Treatment

Among aforementioned neurological diseases, epilepsy constitutes the essential application covered in this book. Epilepsy is a chronic neurological disorder which affects the patients with recurrent seizures. Initiation of seizures is briefly characterized by the excessive electrical discharges in neurons. Duration, frequency, and severity of the seizures are subject to change from patient to patient. Consequently, the outcomes of the seizures may range from involuntary shaking for a couple of seconds to loss of consciousness.

Epilepsy is considered as the most common serious brain disorder throughout the world by World Health Organization (WHO). More explicitly, around 50 million people in the world suffer from epilepsy regardless of age, sex, and location [8]. Although anticonvulsant drug treatment is successful for the majority of the epilepsy patients, the patients that do not respond to drug treatment may undergo a surgery [5].

Briefly, all types of the epilepsy cause additional brain activities. Depending on the place of the activity, the activities can be analyzed in two groups: focal epilepsy, which starts from a certain focus point and propagates, and general epilepsy, in which the entire brain is epileptic. It is expected that focal epilepsy patients could benefit from the conclusion of the research conducted within this framework.

Although majority of the patients suffering from epilepsy could be treated with drug therapy, the patients who do not respond to the drug therapy have to undergo a surgical treatment. The surgical operation aims to first locate and then, resect (or remove) the part(s) of the cortex where epileptic seizures originate and start to propagate.

The surgical treatment is composed of two successive surgeries and a monitoring period in-between: Firstly, the skull is opened and an electrode array is implanted to the pre-located portion of brain. Pre-location is performed by means of magnetic resonance imaging (MRI) prior to the surgery. Subsequently, the skull is closed and the patient is monitored for approximately 2 weeks to locate the focus of epileptic seizures with better precision. In this duration, the neural activities are recorded by an invasive method, namely electrocorticography (ECoG), until the onset of the next seizure. Finally, the second surgery is performed to resect the corresponding brain portion according to the results obtained from ECoG. It is worth noting that epileptogenic tissues are, of course, controlled prior to the removal by using stimulators to ensure that they do not contain critical brain functions. This check takes the mapping of the cortical functional areas into account. Therefore, success and efficiency of the operation can be correlated with the precision of the localization of the epileptic focus.

During an epilepsy surgery, it is aimed to treat the patient by removing the least possible amount of neurons from the brain. Therefore, fine localization of the epileptogenic tissues are critical. The ultimate goal can be defined as single neuronal activity detection if the activity is initiated by a certain neuron. More practically, it is, instead, targeted to localize the region, especially taking the precision of removal tools into account. Therefore, we aim to reduce the area of that region as much as possible. Considering that signal acquisition in EEG is done extracranially, each electrode collects a signal which is an integration of signals coming from thousands of neurons and even worse non-uniformly attenuated versions of these signals. Consequently, intracranial EEG (iEEG) is proposed to improve the attenuation problem and reduce the number of signals, and hence number of neurons, which constitutes the acquired signals on the iEEG electrode. The idea is realized, briefly, by taking the electrodes inside the crane which eliminates the attenuation of signals through the scalp and defines the area of signal integration firmly.

1.2 Motivation and Research Objectives

Next generation neural recording systems [9] aim to alter two main features of conventional iEEG in order to improve the current methodology:

1. Macro iEEG electrodes will be replaced with microelectrode arrays (MEA) fabricated with microtechnology, and
2. transcutaneous wires carrying neural information will be eliminated; thanks to wireless data communication.

Details of the anticipated improvements and benefits of these systems are given in the next section.

1.2.1 Next Generation Neural Recording Systems

The demand from neural recording systems increases continuously in terms of quality and quantity of extracted information with the improvements in the microsystems and microelectronics. This demand drives the technology from external recording systems to in-vivo recording systems. Implanted neural recording systems are expected to offer better spatial and temporal resolution; thanks to implantable microelectrodes and on-site processing microelectronics.

With the introduction of micro electrode arrays (MEAs), first of all, an improvement in the spatial resolution is anticipated [9]. Currently, macro electrodes having a diameter around 10 mm are placed with 10 mm separation on the cortex [4]. However, each unit of MEAs can be fabricated with a diameter less than 100 μm which enhances the precision of localization [10]. For instance, the benefit of this enhancement can be better understood if one considers the epilepsy surgery, where epileptogenic area is, as aforementioned, removed (resected). It is obvious that higher spatial resolution will result in reduced volumes to be removed, and therefore reduce the risk of critical function loss. At this point, it is worth noting that before the resective surgery, the region of interest is also verified not to possess vital functions. Furthermore, in terms of signal characteristics, large electrodes collect an integration of neural activities in the vicinity of the electrode and the result is far from representing single neuron activity. Therefore, miniaturization is anticipated to enable monitoring of single-neuron activity, as well as recording of higher frequency oscillations which cannot be recorded with conventional macro electrodes. Furthermore, utilization of active micro electrodes which also possess higher signal-to-noise ratio (SNR) may enable a better analysis of the epilepsy; thanks to the improved temporal resolution.

Employing transcutaneous wires to transfer neural signals have certain drawbacks both from electrical engineering and medical perspectives. Passive macro electrodes acquire analog signals and transmit them through wires to the external base station where amplifiers and analog-to-digital converters are utilized. However, this method reduces the quality of the raw information acquired from the neurons. Moreover, the wires coming out of the scalp increases the risk of cerebrospinal fluid leakage (CSF), even risk of infection of CSF which may have fatal consequences, and also reduces the patient mobility and comfort. The latter issues may seem less important with respect to former medical risks; however, they, in fact, shorten the monitoring time. Besides, wired connections do not allow life-long implantation which is indispensable for recovering a body function. Wireless data transfer from the implant to the external unit, therefore, offers solid improvements for intracranial recording systems.

On top of these improvements, the signal quality can be further improved by employing on-site low-noise neural recording amplifiers and analog-to-digital converters. Consequently, acquired analog signals are transmitted using digital communication schemes which are more resistant to noise. Utilization of these circuits leads to two new issues, and hence, research questions:

1. Power demand in the implant, and
2. Integration of electrodes and the circuits

Former issue can be overcome by using following solutions or a combination of them:

- Using medical grade batteries,
- Ambient energy harvesting,
- Wireless power transfer.

The application, in fact, determines the solution. For instance, batteries should either be replaced (which requires additional surgery) or recharged for long-term implants. Size restrictions of the implant should also be considered while selecting the battery, as well as the charge capacities of the batteries. Note that recharging an implanted battery requires wireless power transfer. Ambient energy harvesters, on the other hand, may work for ultra-low power implants [11].

Since implantable batteries cannot meet with both the size and capacity requirements simultaneously. Contemporary ambient energy harvesters are not sufficient for supplying several milliwatts of power continuously. Wireless power transfer solutions are therefore appropriate for such long-term in-vivo monitoring of neural signals. Among several candidates in wireless power transfer such as ultrasound, electromagnetic radiation, and magnetic coupling; magnetic coupling provides the most efficient power transfer in short distances in the order of 10–15 mm. Moreover, commonly used frequency range (1–10 MHz [12, 13]) for this method allows sufficient bandwidth for data communication for a reasonable number of electrodes and level of accuracy. It must be noted that any design solution must have a low-power implementation. The aim is to minimize temperature elevation in the surrounding tissues as much as possible.

Compact integration of the implant is critical when size restrictions are considered, thats why, integrated circuit technology is widely exploited in the realization of the implant electronics. Within this technology lies the solution. Integration issue can also be solved by either fabricating the MEAs in CMOS (along with the integrated circuits) or fabricating the MEAs with post-CMOS process. Therefore, the electrodes and the circuits will be inherently integrated.

Following the realization of the entire system, implantation requires one more step: biocompatible packaging. The system should be encapsulated such that it performs all the functions properly and gets accepted by the body. Note that the body will recognize the implanted bare silicon recording system as a threat, so immune system will react against it. However, a biocompatible packaging acts as a barrier to prevent diffusion and consequent chemical reactions in both directions and as an interface which is tolerated by immune system [14, 15].

1.2.2 Research Objectives

The main research objectives addressing the problems defined in this work can be classified under three major subjects:

- Remote powering of active neural recording electrodes and their peripheral electronics
- Wireless transmission of the recorded signals to an external base station for further processing
- Biocompatible packaging of the entire system

Before going further to the challenges and solutions in the wireless power transfer and data communication system, it may be more appropriate to give a background on neural data acquisition get a better grasp of the entire system and cover all the parts of a neural recording system. It will also give us insight to anticipate what kind of challenges are to be expected in the design of such systems.

1.3 Neural Data Acquisition

This section introduces the fundamental concepts of neural data acquisition by means of intracranial electroencephalography (iEEG) with an emphasis on in-vivo epilepsy monitoring. Although clinical scalp EEG is non-invasive, iEEG has become more popular recently since it enables recording of the higher frequency oscillations (HFOs) which give an insight in the electrical mechanisms underlying epilepsy [16].

Acquisition of neural signals is performed by means of micro electrodes which constitute the interface between silicon microelectronics and neurons. These micro electrodes (Fig. 1.2) and their electronic read-out circuits have evolved according to the implant positioning and targeted neural activity. Needle-shaped micro electrodes are preferred for brain-computer interfaces (BCI) in [17] while flat electrodes are employed for cortical surface recordings in [18]. Typical neural recording activities arise from the action potential of a single neuron and local field potential resulting from multi-unit activity. Moreover, the frequency spectrum of the neural activity determines the proper recording electrode type, and hence, the specifications of the electronic circuits. Types of the neural activities, which are critical for epilepsy diagnosis and monitoring, and corresponding frequency bands are presented in Fig. 1.3 according to [16]. Ripples and fast ripples recorded by iEEG may be a valuable indicator for localization of epileptogenic tissues [19–21].

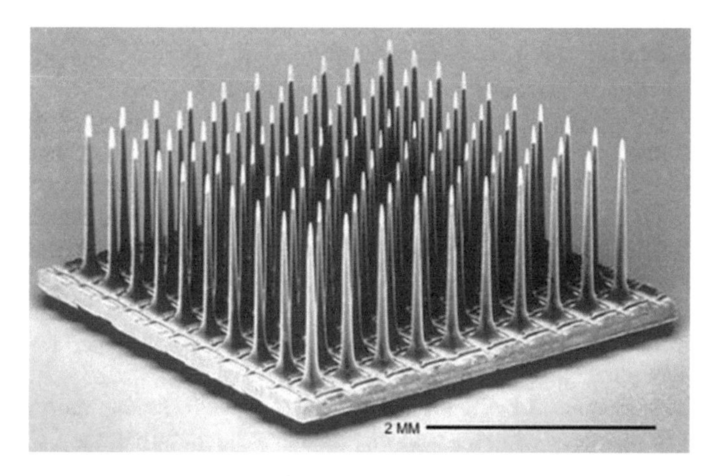

Fig. 1.2 Scanning electron micrograph of silicon-based Utah electrode array [2]

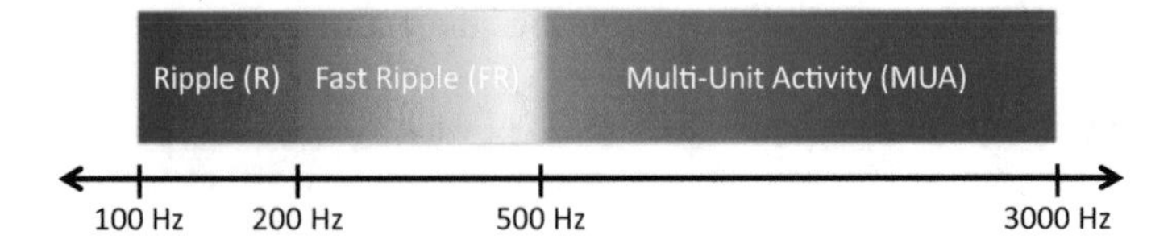

Fig. 1.3 Frequency bands of neural activities [22]

Neural activities are recorded via electrodes by measuring extracellular activities. Although intracellular potential differences can go up to 100 mV, extracellular potential differences have amplitude in the order of hundreds of microvolts. Moreover, the signals attenuate as moved away from the neuron. Therefore in order to achieve better signal amplitude, in other words higher signal-to-noise ratio (SNR) values, the recording electrodes should be placed as close as possible to the cell membrane. On the other hand, electrode impedance and firing of neurons in the vicinity of the electrode contribute to the noise amplitude [23]. Electrode impedance is also critical for high frequency signal acquisition since it degrades high frequency response of the electrodes when combined with distributed capacitance between the neural recording amplifier and the electrode [24]. Consequently, impedance characterization of recording electrodes is critical to evaluate their performances [25]. More specifically, the impedance of neural recording electrodes at 1 kHz is accepted as a benchmark to compare them. In order to reduce the impedance of microelectrode arrays, various materials and processes have been suggested in the last decade. To name a few are carbon nanotube coatings [26] and conducting polymers such as poly(3,4-ethylenedioxythiophene) (PEDOT) [27].

Geometric optimization of these electrodes has to be considered in conjunction with the application. For instance, electrode size should be matched with neuron size for single neuron activity measurements. Furthermore, electrical modeling of electrode-cell interface is quite beneficial while designing the neural amplifiers. Joye et al. [28] presents both point-contact model and area contact model to optimize the electrode diameter for neural recordings. Recently, the demand for high density microelectrode arrays for epilepsy monitoring necessitates a comprehensive study on geometric optimization of electrode size and distance between electrodes.

Following the acquisition of neural signals by MEAs , weak neural signals (10–500 μV) [29] are first amplified with a low-noise amplifier (LNA) (or sometimes called as pre-amplifier), and then digitized with an analog-to-digital converter (ADC). As mentioned previously, the type of activity and the electrode characteristics determines the specifications of these circuits, so over-design should be avoided to minimize power consumption and chip area [30].

As the name implies, LNA amplifies the neural signals with minimum signal-to-noise ratio (SNR) degradation. These circuits have to carefully trade-off two conflicting requirements of being low-power while maintaining a high SNR. Moreover, noise levels vary depending on the bandwidth of neural activity of interest (Fig. 1.3). Therefore, the term noise-efficiency factor (NEF) is coined by Steyaert et al. [31] to compare power and noise performances of amplifiers with respect to a single bipolar transistor:

$$NEF = V_{rms,in} \sqrt{\frac{2\ I_{total}}{\pi U_T 4kT BW}} \tag{1.1}$$

where $V_{rms,in}$ is the total equivalent input noise, I_{total} is the total current drain in the system, and BW is the bandwidth of the amplifier. In the last decade, there have been a great interest in the design of low-power; low-noise amplifier for neural recording [29, 32, 33] following the improvement in NEF presented in [34] which minimizes the input-referred noise by operating the input differential pair in weak inversion and the current mirrors in strong inversion.

Amplified neural signals are digitized by means of an analog-to-digital converter (ADC) to enable digital signal processing on site and digital data communication. Again, low-power operation is the main criterion while choosing the method among several ADC realizations. Due to its low-power requirement and implementation simplicity, successive approximation register (SAR) ADCs are preferred for neural implant applications. The resolution and sampling rate of the ADC is determined by the neural activity of interest. For instance, neural activities given in Fig. 1.3 have different bandwidth requirements, and hence, different sampling rates. Chae et al. [30] present a comprehensive study to optimize these parameters along with chip area and power consumption. They also cover electrode noise which has to be considered in determining ADC resolution in order to prevent unnecessary power consumption. From system level viewpoint, there are two solutions employed for multiple site neural recording systems:

1. Usage of a dedicated ADC for each channel, at the expense of chip area and power consumption [35]
2. Sharing a single ADC is between the channels through a multiplexer, at the expense of losing information [36]

For high frequency oscillation (HFO) recordings and/or multiple site recordings, the payload of uplink communication, i.e. from the implant to the external base station, grows enormously. In order to minimize power consumption due to data communication, data compression methods can be employed without losing critical information. The majority of the literature on this application relies on delta compression which computes the differences between signals in successive frames and creates an output if the result is higher than a pre-defined threshold [37]. Obviously, this results in a loss of information. However, depending on the sparsity of the signal, the benefit can overcome this loss. For instance, Rizk et al. [38] presents 15-to-1 signal compression for spike sorting which reduces the data payload drastically. Recently, compressive sensing based payload reduction solutions, that rely on the fact that action potentials are sparse, have also been presented [39].

Digitized neural signals from each channel are converted into a serial bit-stream which constitutes the payload of the wireless uplink communication. The conversion enables low-power digital signal processing and high quality data transmission.

Signal processing aims to reduce the amount of data to be transferred, and hence, data rate and required bandwidth. Moreover, low data rate communication results in less power consumption in the implant. Payload of the uplink data communication depends on the following specifications of the neural recording:

- Number of active electrodes at an instant
- Number of recording sites
- Sampling rate of ADC
- Resolution of ADC
- Compression factor

For instance, an in-vivo epilepsy monitoring system which is composed of 5 recording sites and 16 electrodes at each site, samples at 30 kHz (for multi-unit activity), and resolves for 8-bit will require a 19.2 Mbps uplink communication without compression. This calculation, indeed, reveals the importance of compression since the data rate will be reduced by compression factor.

1.4 Anticipated Challenges

Combining the specifications imposed by the acquisition system with the anticipated improvements on the epilepsy surgery and medical regulations, a list of anticipated challenges and trade-offs have been listed such that research effort could be focused on certain subjects:

- Size: First and foremost, the system will be implanted in the brain so an extensive use of microtechnologies is required both at the circuit and system level.
- Temperature elevation: The implant must respect the temperature elevation regulations imposed for implanted medical devices. Therefore, low-power circuit designs along with highly efficient power-transfer for the remote powering have to be realized.
- Data rate: Using multiple number of recording sites with higher sampling rate necessitates a wide bandwidth wireless data communication link.
- Biocompatible packaging: The implant should be packaged appropriately such that it will be accepted by the body and will work properly at the same time. The quality of the packaging should also be compatible with the anticipated duration of 2 weeks of intracranial monitoring.

Next chapter proposes solutions at the system level to address the aforementioned challenges.

1.5 Book Outline

Contents of this chapter have been partially published in [40]. In Chap. 2, the system level approaches to build a wireless system for two different target applications have been introduced. Chapter 3 explains the wireless power transfer part of the overall system with a detailed circuit and system analysis. Chapter 4 introduces the details of the implemented bidirectional communication link using two different methods. In Chap. 5, biocompatible packaging of the implant has been detailed; along with temperature requirements for intra-cranial neural implants. Chapter 6 gives the complete experimental results of the implemented system in various environments, including a partial implantation in a rat brain.

References

1. G. Deuschl, C. Schade-Brittinger, P. Krack, J. Volkmann, H. Schfer, K. Btzel, C. Daniels, A. Deutschlnder, U. Dillmann, W. Eisner, D. Gruber, W. Hamel, J. Herzog, R. Hilker, S. Klebe, M. Klo, J. Koy, M. Krause, A. Kupsch, D. Lorenz, S. Lorenzl, H.M. Mehdorn, J. R. Moringlane, W. Oertel, M.O. Pinsker, H. Reichmann, A. Reu, G.-H. Schneider, A. Schnitzler, U. Steude, V. Sturm, L. Timmermann, V. Tronnier, T. Trottenberg, L. Wojtecki, E. Wolf, W. Poewe, J. Voges, A randomized trial of deep-brain stimulation for parkinson's disease. New England J. Med. **355**(9), 896–908 (2006)
2. R.R. Harrison, The design of integrated circuits to observe brain activity. Proc. IEEE **96**(7), 1203–1216 (2008)
3. W.H. Theodore, R.S. Fisher, Brain stimulation for epilepsy. Lancet Neurol. **3**(2), 111–118 (2004)
4. J.J. Van Gompel, S.M. Stead, C. Giannini, F.B. Meyer, W.R. Marsh, T. Fountain, E. So, A. Cohen-Gadol, K.H. Lee, G.A. Worrell, Phase i trial: safety and feasibility of intracranial electroencephalography using hybrid subdural electrodes containing macro- and microelectrode arrays. Neurosurg. Focus **25**(3), E23 (2008)
5. J.F. Tllez-Zenteno, R. Dhar, S. Wiebe, Long-term seizure outcomes following epilepsy surgery: a systematic review and meta-analysis. Brain **128**(5), 1188–1198 (2005)
6. E. Carrette, K. Vonck, V. De Herdt, A. Van Dycke, R. El Tahry, A. Meurs, R. Raedt, L. Goossens, M. Van Zandijcke, G. Van Maele, V. Thadani, W. Wadman, D. Van Roost, P. Boon, Predictive factors for outcome of invasive video-eeg monitoring and subsequent resective surgery in patients with refractory epilepsy. Clin. Neurol. Neurosurg. **112**(2), 118–126 (2010)
7. H.M. Hamer, H.H. Morris, E.J. Mascha, M.T. Karafa, W.E. Bingaman, M.D. Bej, R.C. Burgess, D.S. Dinner, N.R. Foldvary, J.F. Hahn, P. Kotagal, I. Najm, E. Wyllie, H.O. Lders, Complications of invasive video-EEG monitoring with subdural grid electrodes. Neurology **58**(1), 97–103 (2002)
8. R.A. Scott, S.D. Lhatoo, J.W. Sander, The treatment of epilepsy in developing countries: where do we go from here? Bull. World Health Organ. **79**(4), 344–351 (2001)
9. A. Yakovlev, S. Kim, A. Poon, Implantable biomedical devices: wireless powering and communication. IEEE Commun. Mag. **50**(4), 152–159 (2012)
10. M. Shoaran, C. Pollo, Y. Leblebici, A Schmid, Design techniques and analysis of high-resolution neural recording systems targeting epilepsy focus localization, in *2012 Annual International Conference of the IEEE Engineering in Medicine and Biology Society (EMBC)*, pp. 5150–5153 (2012)
11. P.P. Mercier, A.C. Lysaght, S. Bandyopadhyay, A.P. Chandrakasan, K.M. Stankovic, Energy extraction from the biologic battery in the inner ear. Nat. Biotech. **30**(12), 1240–1243 (2012)

12. R. Harrison, P. Watkins, R. Kier, R. Lovejoy, D. Black, R. Normann, F. Solzbacher. A low-power integrated circuit for a wireless 100-electrode neural recording system, in *Digest of Technical Papers IEEE International Solid-State Circuits Conference (ISSCC 2006)*, pp. 2258–2267 (2006)
13. C.M. Lopez, D. Prodanov, D. Braeken, I. Gligorijevic, W. Eberle, C. Bartic, R. Puers, G. Gielen, A multichannel integrated circuit for electrical recording of neural activity, with independent channel programmability. IEEE Trans. Biomed. Circuits Syst. **6**(2), 101–110 (2012)
14. C. Hassler, T. Boretius, T. Stieglitz, Polymers for neural implants. J. Polym. Sci. Part B: Polym. Phys. **49**(1), 18–33 (2011)
15. T. Stieglitz, Manufacturing, assembling and packaging of miniaturized neural implants. Microsyst. Technol. **16**(5), 723–734 (2010)
16. C.A. Schevon, A.J. Trevelyan, C.E. Schroeder, R.R. Goodman, G. McKhann, R.G. Emerson, Spatial characterization of interictal high frequency oscillations in epileptic neocortex. Brain **132**(11), 3047–3059 (2009)
17. E.M. Maynard, C.T. Nordhausen, R.A. Normann, The utah intracortical electrode array: a recording structure for potential brain-computer interfaces. Electroencephalogr. Clin. Neurophysiol. **102**(3), 228–239 (1997)
18. D. Yoshor, William H. Bosking, Geoffrey M. Ghose, H.R. J. Maunsell, Receptive fields in human visual cortex mapped with surface electrodes. Cereb. Cortex **17**(10), 2293–2302 (2007)
19. T. Akiyama, B. McCoy, C.Y. Go, A. Ochi, I.M. Elliott, M. Akiyama, E.J. Donner, S.K. Weiss, O. Carter Snead, J.T. Rutka, J.M. Drake, H. Otsubo, Focal resection of fast ripples on extraoperative intracranial eeg improves seizure outcome in pediatric epilepsy. Epilepsia **52**(10), 1802–1811 (2011)
20. A. Bragin, I. Mody, C.L. Wilson, J. Engel, Local generation of fast ripples in epileptic brain. J. Neurosci. **22**(5), 2012–2021 (2002)
21. S. Wang, I.Z. Wang, J.C. Bulacio, J.C. Mosher, J. Gonzalez-Martinez, A.V. Alexopoulos, I.M. Najm, K.S. Norman, Ripple classification helps to localize the seizure-onset zone in neocortical epilepsy. Epilepsia **54**(2), 370–376 (2013)
22. G. Yilmaz, C. Dehollain, Single frequency wireless power transfer and full-duplex communication system for intracranial epilepsy monitoring. Microelectron. J. **45**(12), 1583–1834 (2014)
23. Z. Yang, Q. Zhao, E. Keefer, W. Liu, Noise characterization, modeling, and reduction for in vivo neural recording, in ed. by Y. Bengio, D. Schuurmans, J. Lafferty, C.K.I. Williams, A. Culotta *Advances in Neural Information Processing Systems 22*, pp. 2160–2168 (2009)
24. D.A. Robinson, The electrical properties of metal microelectrodes. Proc. IEEE **56**(6), 1065–1071 (1968)
25. S.F. Cogan, Neural stimulation and recording electrodes. Ann. Rev. Biomed. Eng. **10**(1), 275–309 (2008)
26. E.W. Keefer, B.R. Botterman, M.I. Romero, A.F. Rossi, G.W. Gross, Carbon nanotube coating improves neuronal recordings. Nat Nano **3**(7), 434–439 (2008)
27. K.A. Ludwig, N.B. Langhals, M.D. Joseph, S.M. Richardson-Burns, J.L. Hendricks, D.R. Kipke, Poly(3,4-ethylenedioxythiophene) (pedot) polymer coatings facilitate smaller neural recording electrodes. J. Neural Eng. **8**(1), 014001 (2011)
28. N. Joye, A. Schmid, Y. Leblebici, Electrical modeling of the cellelectrode interface for recording neural activity from high-density microelectrode arrays. *Neurocomputing*, 73(13):250–259, 2009. Timely Developments in Applied Neural Computing (EANN 2007)/Some Novel Analysis and Learning Methods for Neural Networks (ISNN 2008)/Pattern Recognition in Graphical Domains
29. W. Wattanapanitch, M. Fee, R. Sarpeshkar, An energy-efficient micropower neural recording amplifier. IEEE Trans. Biomed. Circuits Syst. **1**(2), 136–147 (2007)
30. S.C. Moo, L. Wentai, M. Sivaprakasam, Design optimization for integrated neural recording systems. IEEE J. Solid-State Circuits **43**(9), 1931–1939 (2008)
31. M.S.J. Steyaert, W.M.C. Sansen, A micropower low-noise monolithic instrumentation amplifier for medical purposes. IEEE J. Solid-State Circuits **22**(6), 1163–1168 (1987)

32. J. Holleman, B. Otis, A sub-microwatt low-noise amplifier for neural recording, in *29th Annual International Conference of the IEEE Engineering in Medicine and Biology Society (EMBS 2007)*, pp. 3930–3933 (2007)
33. V. Majidzadeh, A. Schmid, Y. Leblebici, Energy efficient low-noise neural recording amplifier with enhanced noise efficiency factor. IEEE Trans. Biomed. Circuits Syst. **5**(3), 262–271 (2011)
34. R.R. Harrison, C. Charles, A low-power low-noise cmos amplifier for neural recording applications. IEEE J. Solid-State Circuits **38**(6), 958–965 (2003)
35. M. Mollazadeh, K. Murari, G. Cauwenberghs, N. Thakor, Micropower cmos integrated low-noise amplification, filtering, and digitization of multimodal neuropotentials. IEEE Trans. Biomed. Circuits Syst. **3**(1), 1–10 (2009)
36. R.R. Harrison, P.T. Watkins, R.J. Kier, R.O. Lovejoy, D.J. Black, B. Greger, F. Solzbacher, A low-power integrated circuit for a wireless 100-electrode neural recording system. IEEE J. Solid-State Circuits **42**(1), 123–133 (2007)
37. J.N.Y. Aziz, K. Abdelhalim, R. Shulyzki, R. Genov, B.L. Bardakjian, M. Derchansky, D. Serletis, P.L. Carlen, 256-channel neural recording and delta compression microsystem with 3d electrodes. IEEE J. Solid-State Circuits **44**(3), 995–1005 (2009)
38. M. Rizk, I. Obeid, S.H. Callender, P.D. Wolf, A single-chip signal processing and telemetry engine for an implantable 96-channel neural data acquisition system. J. Neural Eng. **4**(3), 309 (2007)
39. Z. Charbiwala, V. Karkare, S. Gibson, D. Markovic, M.B. Srivastava, Compressive sensing of neural action potentials using a learned union of supports, in *2011 International Conference on Body Sensor Networks (BSN)*, pp. 53–58 (2011)
40. G. Yilmaz, C. Dehollain, Intracranial epilepsy monitoring using wireless neural recording systems, in ed. by S. Carrara, K. Iniewski *Handbook of Bioelectronics: Directly Interfacing Electronics and Biological Systems*, vol. 008, pp. 389–399. Cambridge University Press, Cambridge (2015)

Chapter 2
System Overview

Abstract Upon giving an insight for the next-generation neural recording systems, this chapter introduces the proposed solutions for wireless power transfer and wireless data communication for intracranial neural implants in the scope of this book, i.e., epilepsy monitoring. First of all, a system overview is given in terms of functional blocks and the methods anticipated to be used for implementation. Next, a brief explanation of each function is presented while leaving the detailed analysis for following chapters. Moreover, a summary of the system specifications is given in order to complete the system-level idea.

2.1 System Specifications

As aforementioned in Sect. 1.3, fast ripples (FR) have already been correlated to be a strong indicator for the localization of epileptogenic zones, and multi-unit activity (MUA) signals are anticipated to reveal more about the nature of the epilepsy and recently widely explored in the literature.

Therefore, in the scope of this book, two applications that aim to record FR and MUA signals are targeted. Since the characteristics of these signals are different, in particular in terms of frequency spectrum and hence interface electronics, two sets of specifications corresponding to target applications have been defined. Table 2.1 summarizes the system specifications targeted as the case study of this book.

The targeted data rate for fast ripple recording system which provides significant information for epilepsy monitoring has been calculated as follows: Fast ripple activity covers a frequency range from 200 to 500 Hz [1], which has to be considered while designing the neural recording amplifier [2] and the analog-to-digital converter (ADC). Consequently, the sampling rate for fast ripple (<500 Hz) detection has been set as 2 kSample/s. Using an analog-to-digital converter with a 8-bit resolution for 64 recording channels (4 recording sites and each site contains 16 electrodes) required

G. Yılmaz and C. Dehollain, *Wireless Power Transfer and Data Communication for Neural Implants*, Analog Circuits and Signal Processing,
DOI 10.1007/978-3-319-49337-4_2

Table 2.1 System specifications for FR and MUA signal recording applications

	Fast ripple recording	Multi-unit activity recording
Number of recording sites	4	4
Number of electrodes per site	16	4
Bandwidth of signal of interest (Hz)	200–500	500–3000
ADC sampling rate (Sample/s)	2000	30,000
ADC resolution	8-bit	8-bit
Anticipated compression factor	0.5	0.5
Total data rate budget (kbps)	512	1920
Total power budget (mW)	9.6	9.6

data rate is calculated to be 512 kbps with an on-site compression factor of 0.5. More explicitly, total data rate budget is calculated by multiplying:

- Number of recording sites
- Number of electrodes per site
- ADC sampling rate
- ADC resolution
- Compression factor

Same calculation applies for multi-unit activity recording with its corresponding numbers shown in Table 2.1. Particularly for multi-unit activity recording, the sampling specifications are defined to match with state-of-the-art high-end iEEG monitoring devices that are already exploited in clinical use.

The number of electrode sites and the number of electrodes per site have also been defined by neurosurgeons specialized in epilepsy surgeries. A power budget estimation of 2.4 mW per recording site has been made.

In addition to Table 2.1, a downlink data rate of 1 kbps is aimed to be able to send commands to the implant in order to control certain parameters, such as selection of active sites and configuration of acquisition parameters. In fact, 1 kbps is already more than required for such tasks.

2.2 System-Level Solutions

Figure 2.1 illustrates the implantation idea for the proposed intracranial neural recording system. The implant part of the entire system is composed of a master radio frequency (RF) unit, which is in fact the core material of this book, and its satellite sensor chips which contain microelectrode arrays and their interface circuits such as amplifiers and ADCs. Master RF unit and sensors will be connected with flexible

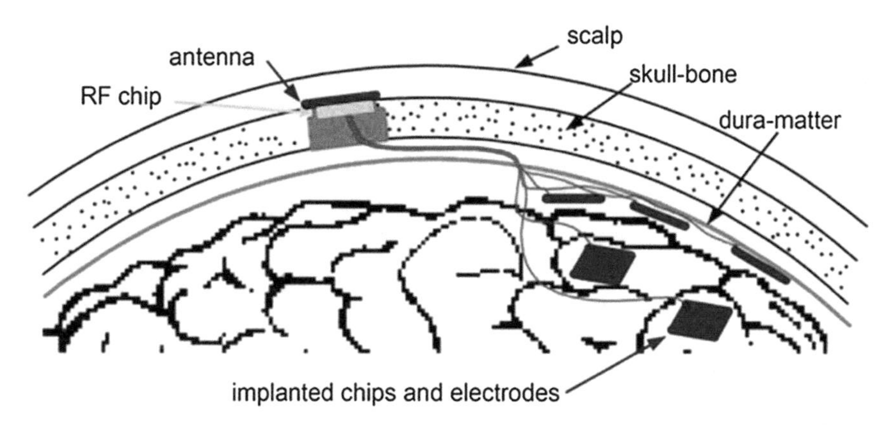

Fig. 2.1 Schematics of the implanted system; here, five recording-integrated circuits are depicted on the surface of the cortex, in *dark blue*; these units are connected through a wire (*red*) to the RF chip (*orange*), which is located inside a package mounted into the Burr hole

wires which will carry power and data to the sensors as well as digitized neural data from the sensor chips to the master RF unit which will transmit it to the external base station. The external base station will also be responsible from transferring power and data to the master RF unit. Novelty of this system is using distributed microelectrode arrays (MEAs) instead of using a single MEA. Consequently, it is anticipated that more information could be extracted thanks to the increased coverage area on the cortex.

The master RF unit, containing wireless power transfer and data communication system, is envisaged to be implanted in the Burr hole which is opened on the skull for neurosurgical treatment of epilepsy. This hole can be defined as a cylinder having a height of approximately 10 mm (average skull thickness) and a diameter of 15 mm (subject to change depending on the drill size). These numbers, in fact, determine the size limitations for the system proposed in this book.

Returning back to the target applications, two different solutions are proposed to address them and satisfy the specifications in Table 2.1. These solutions differ from each other with the approach employed for uplink communication. More explicitly, a single-frequency approach, in which wireless power transfer and bidirectional data communication are performed at the same frequency, is adapted for fast ripple recording application, whereas two-frequency approach, in which uplink data communication is performed at a frequency different from the operation frequency for wireless power transfer and downlink communication, is utilized for multi-unit activity recording. Briefly, it is due to the fact that carrier frequency selected for wireless power transfer is not suitable for high data rate uplink communication as requested for multi-unit activity. The reasoning behind the selection of power transfer frequency is detailed in Chap. 3.

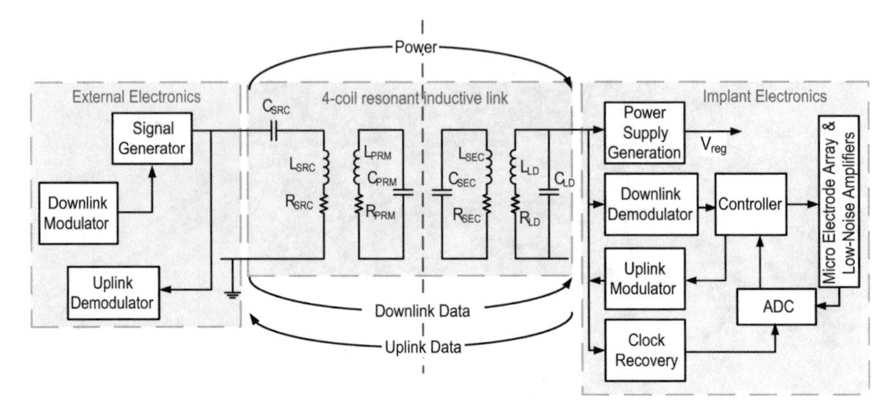

Fig. 2.2 System-level representation of the proposed single-frequency system composed of external electronics, 4-coil resonant inductive link, and implant electronics [1]

2.2.1 *Single-Frequency Approach*

As mentioned in Sect. 1.2, main challenges to be handled are wireless power transmission to the implant, bidirectional communication, and biocompatible packaging of the system. This section deals with the questions related to electronics and hence is divided into three main functions: wireless power transfer, downlink communication, and uplink communication.

Figure 2.2 depicts the system-level representation of the proposed single-frequency approach with its functional blocks. Although at first glance one may think that separating power and data transfer channels enables high performance in both, we may be forced to optimize the performance of both power and data links on the same physical link due to the size limitations imposed by the implantation process.

2.2.1.1 Wireless Power Transfer

Remote powering, basically, addresses the goal of transmitting energy from an external power source to the sensors. Note that high-performance sensors designed recently are generally active, and hence, they dissipate power. Moreover, most of these sensors require peripheral electronics to start up, operate, and turn off. Although the power dissipation of the circuits is not comparable with the sensors thanks to the state-of-the-art IC technologies, they are still not ignorable since they determine the standby power requirement of the implanted systems for in vivo sensing.

Remote powering solutions can be analyzed in two sections with respect to the distance between the external base station and the implanted unit: near field and far field. These definitions have already been done according to the value of the distance in comparison with the wavelength of the source signal.

Long distance (a few meters) remote powering solutions generally exploit far-field properties, more explicitly radiation properties, of antennas at several hundreds of MHz frequencies. Therefore, they are more suitable for applications which require high mobility. On the other hand, short distance (a few centimeters) remote powering solutions employ reactive coupling techniques such as capacitive and inductive at several MHz frequencies. As the names imply, capacitive coupling is a result of electrical coupling, whereas inductive coupling is of magnetic coupling. Capacitive coupling requires a dielectric medium that allows strong coupling and is more sensitive to distance variations. On the other hand, inductive coupling method exploits the mutual inductance between coupled inductors [3], so it is more preferable for powering the implanted devices.

It is clear that efficient energy transfer from the base station to the implanted unit is critical considering the medical issues due to overexposure of power in terms of both amplitude and time. On the other hand, noting that the base station is also portable and powered by a battery, energy transfer efficiency becomes significant in terms of system-level design. Since the total efficiency can be expressed as the multiplication of efficiency of every single unit, improving the least efficient unit is the most crucial task. Analysis of literature in this perspective shows that the weakest link of the chain is the inductive link [3]. Consequently, in this system, 4-coil resonant inductive link topology is envisaged to be utilized thanks to its superior power transfer efficiency compared to other topologies. Details of this selection is given in Chap. 3.

Realization of the inductive coupling in system level requires a signal source and a load which is actually the sensors and electronics of the implanted biomedical device. Due to practical reasons, the signal source is realized with a signal generator and sometimes fortified with a driver and a power amplifier. Moreover, the load of the inductive link begins with a power management unit, which is composed of a rectifier and a voltage regulator, followed by integrated electronics, and ends with sensors and/or actuators.

2.2.1.2 Downlink Communication

It is possible to cover a larger area of the cortex by implanting multiple MEAs which is controlled by a central unit. Therefore, a communication channel from the external world to the implant (downlink) is required to select and configure MEAs. Moreover, as detailed in Fig. 1.3, neural activities have different bandwidths which thereby necessitate different sampling rates for on-site digitization. Note that acquired neural signals via MEAs are first amplified with low-noise amplifiers (LNAs) and then digitized by analog-to-digital converters (ADCs). Consequently, it is clear that sending configuration data to the implant will yield smarter solutions in terms of efficiency.

Therefore, a downlink communication channel will be established on the wireless power transfer link by modulating the amplitude of the source signal supplied by the signal generator. This variation will be detected using an amplitude-shift-keying (ASK) demodulator on the implant side as illustrated in Fig. 2.2.

2.2.1.3 Uplink Communication

A communication channel from the implant to an external base station (uplink) is required to transmit the digitized neural data to the external world. In the single-frequency approach, uplink communication is also performed on the wireless power transfer channel by exploiting the idea of load modulation used in RFID systems [4]. Briefly, changing a parameter on the implant side will create a change on the external base station side since they are mutually coupled due to the resonant inductive link. Creating this variation with the digitized neural data will mean that neural information can be transferred to external base station as long as the variation could be detected. Therefore, a modulator, which is composed of a switch and a detuning capacitor, is envisaged to be implemented in the implant side and an ASK demodulator on the external base station to establish the uplink communication channel.

2.2.2 Two-Frequency Approach

In order to satisfy the increasing data rate requirements for the multi-unit activity recording applications, uplink communication channel is established at a higher frequency than the power transfer frequency in order to increase the available bandwidth. In fact, the wireless power transfer and downlink communication solutions are exactly the same as the ones for the single-frequency approach. Therefore, this section solely explains the operation principle of the uplink communication. Figure 2.3 illustrates the two-frequency approach to be used for higher data rate uplink communication.

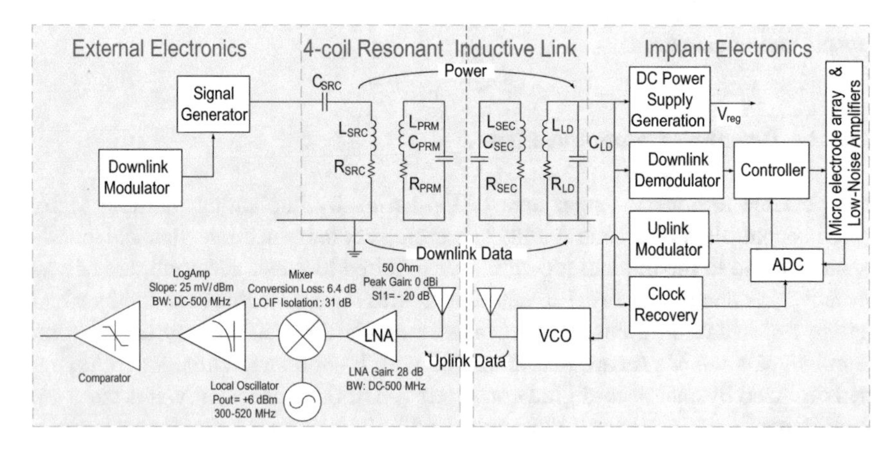

Fig. 2.3 System-level representation of the two-frequency approach which performs wireless power transfer and downlink communication at the same frequency, whereas uses a higher frequency for uplink communication

2.2.2.1 Uplink Communication

For the two-frequency approach, an uplink communication channel at a higher frequency is proposed in order to have sufficient bandwidth and to satisfy the data rate requirements of multi-unit activity recording. The transmitter in the implant will be composed of a voltage-controlled oscillator (VCO) and an antenna for radiation. VCO will be modulated with the digitized neural information, and radiated signals will be received by the receiver antenna on the external base station. Afterward, the received signal will be amplified and downconverted using a mixer and a local oscillator to near-zero IF band. Next, mixer output will be amplified with a logarithmic amplifier and finally turned into a rail-to-rail digital waveform using a high-speed comparator.

After explaining the uplink communication solution using a dedicated transmitter, a short discussion on how to choose between these two methods can be given. In the single-frequency approach, detuning of the resonance frequency is proposed using an additional detuning capacitance. Note that the amount of detuning has to be increased with increasing data rate to achieve same energy-per-bit value (E_b) or bit-error-rate (BER). This actually sets the decision point whether to use the single-frequency or two-frequency approach since wireless power transfer efficiency decreases as amount of detuning increases as will be explained in Chap. 4. If a dedicated transmitter, which does not violate the size requirements, can transmit the same amount of data while keeping the BER level same as with its load modulation version at a lower power consumption, it is, of course, more logical to shift from the single-frequency solution to two-frequency approach.

2.3 Summary

In this chapter, system-level approach to the defined problem with intracranial epilepsy monitoring is addressed. In the frame of epilepsy research, two applications targeting to record different neural signals, namely as fast ripples and multi-unit activity, have been defined and system-level specifications for these applications have been set. Moreover, solutions for wireless power transfer and bidirectional data communication have been proposed to cover these two applications. Following two chapters give more detailed information regarding these solutions.

References

1. C.A. Schevon, A.J. Trevelyan, C.E. Schroeder, R.R. Goodman, G. McKhann, R.G. Emerson, Spatial characterization of interictal high frequency oscillations in epileptic neocortex. Brain **132**(11), 3047–3059 (2009)
2. M. Shoaran, C. Pollo, Y. Leblebici, A. Schmid, Design techniques and analysis of high-resolution neural recording systems targeting epilepsy focus localization, in *2012 Annual International*

Conference of the IEEE Engineering in Medicine and Biology Society (EMBC), pp. 5150–5153 (2012)
3. R.R. Harrison, Designing efficient inductive power links for implantable devices, in *IEEE International Symposium on Circuits and Systems, 2007. ISCAS 2007* (2007), pp. 2080–2083
4. K. Finkenzeller, *RFID Handbook: Fundamentals and Applications in Contactless Smart Cards and Identification*, 2nd edn. (Wiley, New York, 2003)

Chapter 3
Wireless Power Transfer

Abstract This chapter introduces the fundamentals of wireless power transfer with an emphasis on implant powering. Firstly, an overview of implant powering solutions is introduced, and then, the decision of using wireless power transfer is justified. Next, among possible wireless power transfer methods, it is explained that why magnetic coupling befits the best for the target application and its specifications. In this book, a recent inductive link topology which employs 4 coils at resonance is employed in order to realize magnetic coupling. This approach has brought two main advantages: higher power transfer efficiency and less dependence of power transfer efficiency on load impedance. Detailed explanation of the 4-coil resonant inductive link is followed by the design of electronic circuits utilized to create a reliable power supply in the implant. This unit consists of an active half-wave rectifier and a low drop-out voltage regulator.

3.1 Implant Powering Solutions

Addressing the differences between the current clinical practice and the emerging wireless solution for intracranial recording, the necessity of wireless power transmission can be explained more explicitly. In the first stage of the two-step resective epilepsy surgery, an electrode array, which is composed of millimeter-sized recording electrodes, is placed onto the cortex as shown in Fig. 3.1. Cable bundles are used to carry the acquired neural signals to the external processing unit. These bundles carry information of a specific row (or column depending on the convention) passing through the dura mater, skull, and scalp, respectively. Since these recording electrodes are passive and signal processing is done externally, the implanted electronics does not demand any power.

However, eliminating these cables in order to achieve a completely wireless solution necessitates a power supply in the implant, i.e., inside the crane. At first glance, this power demand can be basically attributed to the wireless data transmission; how-

G. Yılmaz and C. Dehollain, *Wireless Power Transfer and Data Communication for Neural Implants*, Analog Circuits and Signal Processing,
DOI 10.1007/978-3-319-49337-4_3

Fig. 3.1 Placement of an electrode array on the cortex during the first stage of the resective epilepsy surgery [1] (courtesy of Sydney S. Cash)

ever, the inclusion of a power supply in the implant which enhances the quality of the signal recording, such as on-site digitization, demands extra power for its operation.

In the literature, the solutions to the power demand in the implant are addressed using either one or a combination of the methods listed below. From a technical point of view, these three methods listed below differ in just how transformation of energy is realized; naturally obeying the first law of thermodynamics:

- ambient energy harvesting
- battery usage
- wireless power transfer (remote powering)

As a matter of fact, all the mentioned methods have advantages and drawbacks for specific applications, and hence, for specifications, restrictions, and regulations imposed for that specific case.

3.1.1 Ambient Energy Harvesting

This method aims to harvest energy from the sources which are already available in the surrounding medium. Possible examples for these energy sources can be listed as light [2, 3], electromagnetic radiation [4, 5], thermal gradients [6–8], and kinetic energy which is widely used in watch industry since decades. In literature, motion, vibration, and kinetic energy are the most common forms of energy that are converted to electrical energy. Mitcheson et al. gives an extensive review of studies focusing on energy harvesting from human and machine motion in [9]. As discussed in the same study, ambient energy harvesters suffer from low-quantity and low-quality power output where low-quality power output refers to varying power levels and even sometimes zero output power. Considering continuous power demand of the neural implants aiming for continuous data transmission and the estimated power budget, current ambient energy harvesters are found to be insufficient to fulfill this task [10]. Nevertheless, it is fair to note that the power demand of electronics are decreasing and power density of ambient energy harvesters are increasing which points out an intersection in the near future.

3.1.2 Battery Usage

Using a battery to power up a system which has no wired connections to an energy source is the most common and the easiest method for numerous mobile applications. Moreover, it is certain that a battery will create a more reliable power supply than an ambient energy harvester. On the other hand, certain regulations have to be satisfied in order to be able to implant a battery inside the body, which can be investigated under biocompatibility. Briefly, batteries transform chemical energy to electrical energy and these chemicals and/or the products of the reactions might be hazardous for the body. In that case, body reacts by showing foreign body reaction to those substances, which causes inflammation and may eventually result in removal of the implant. This obstacle can be overcome by employing medical-grade batteries which have already been approved to be biocompatible. Following the regulations, it is necessary to evaluate these batteries in terms of the capacity to size ratio. Batteries occupy a significant amount of volume for hearing aids and pacemakers; however, the volume allocated for intracranial neural implants is quite small compared to these applications. For instance, lithium-polymer thin-film batteries are promising in terms of dimensions, particularly with a thickness less than 0.2 mm. However, even the largest ones (25 mm $\times$ 50 mm, by Infinite Power Solutions Inc.) can supply an implant with an average power demand of 10 mW only for an hour. At this rate, replacing the battery is not a feasible solution. Consequently, a combined solution that involves a battery and a recharging unit either by ambient energy harvesting or wireless power transmission can be employed as in [3].

3.1.3 Wireless Power Transfer

This method can be explained as a well-structured version of ambient energy harvesters. The difference is that using a dedicated external source, it is guaranteed that the source of energy is always available. Moreover, it is possible to adjust the power level according to varying power demand of the implant. Consequently, the drawbacks of the ambient energy harvesters are eliminated. In the literature, there are numerous examples of wireless power transfer by means of electromagnetic (EM) radiation [11, 12], magnetic coupling [13–15], ultrasonic coupling [16], and infrared radiation [17]. It is fair to claim that EM radiation and magnetic coupling-based systems dominate the literature, especially for neural implant powering applications. This can be attributed to the easiness of manufacturing and integration of inductive links and antennas with electronics in a single research group, while design and manufacturing of ultrasonic transducers and infrared sensors require a different expertise.

Consequently, among these three methods, wireless power transfer stands out as the most appropriate one for implant powering. It can be fortified with an implantable battery, if the final size of the implant can accommodate a battery.

3.2 Wireless Power Transfer

As explained in the previous section, there are different methods of realizing wireless power transfer to power up the implant. Since there is no ideal method which works for any conditions, the target application itself should be analyzed in order to decide which method should be employed. In this case, i.e., a subdural neural implant, there are two major constraints which are, actually, interlinked:

1. the separation distance, composition of the separation, and the allowed implant size and
2. the regulations limiting the power dissipation in order to guarantee a safe operation on the cortex in terms of tissue heating.

Therefore, the chosen method should respect these two limitations regardless of its performance. In the literature, it is quite common to focus only on the power transfer efficiency to evaluate the performance of the system while staying inside the limitations. However, even if the numbers related with these two limitations are clearly expressed, it may be deceiving to evaluate the performance of a WPT system without knowing the parameters tabulated in Table 3.1.

Revisiting the imposed limitations and the targeted specifications, the resulting table for epilepsy monitoring can be formed as in Table 3.2. As indicated in Table 3.2, the form factor of the implant could be defined by using the constraints imposed by the dimensions of the Burr hole, which is the initial hole drilled at the beginning of the epilepsy surgery. The implant has to be fit inside this cylindrical volume. Moreover, temperature limitations have to be respected in order to fulfill the regulations for body implants. Regulations allow 1 °C temperature elevation for body implants. This temperature rise corresponds to 40 mW/cm^2 power outflux density. Since the acquisition chips dissipate 2.4 mW per site, at least 6 mm^2 surface area is required to respect this regulation. Noting that the acquisition chips are occupying an area of 1.5 mm × 3 mm (4.5 mm^2) on silicon substrate, the total surface area can be approximated as 9 mm^2; neglecting quite small surface areas due to silicon chip thickness. Dividing the power consumption per site to the surface area, following power outflux density is reached:

$$2.4\ \mathrm{mW}/9\,\mathrm{mm}^2 = 27\ \mathrm{mW}/\mathrm{cm}^2$$

which corresponds to 0.675 °C temperature elevation. Therefore, it can be concluded that the system is respecting the temperature regulations inside the brain.

Now that possible methods have been introduced and limitations for the case of epilepsy monitoring have been indicated, most suitable wireless power transfer method can be selected. Considering the given medium composition, magnetic coupling, and ultrasonic coupling have an advantage over two other methods, particularly over infrared radiation. Moreover, EM radiation causes higher energy absorption in the tissues which gives rise to the temperature [18]. Another aspect is that using magnetic coupling, energy can be transferred to the sensor nodes more efficiently than EM radiation in small separation distances [19]. Additionally, it is important to differentiate the power losses due to the implementation of the method and the power loss on the tissue to have a fair comparison while staying in the limits of the regulations. As explained in [18], at low MHz frequencies, tissues absorb less power than a few hundreds of MHz, where EM radiation applications take place. Another

Table 3.1 Important parameters for performance evaluation of WPT systems

Parameter	Importance
Open circuit ac voltage created at the output of the system	To turn on the rectifier
Available output power	To assure a proper operation of all the electronics and the electrodes (if active) (in addition to the power dissipated in the implant)
Absorbed power by the tissues	To calculate the temperature elevation in the surrounding tissues more accurately
Source power	To evaluate mobility and autonomy of the external unit

Table 3.2 Summary of limitations and specifications of the target application

Constraints	Specifications
Separation distance	10 mm
Composition of gap	Bone and skin
Allowed implant size	A cylinder with 15 mm diameter and 10 mm height
Temperature regulations	Maximum 1 °C elevation or 40 mW/cm^2 power density
Open circuit ac voltage	>2.3 V (to reach 1.8 V DC)
Available output power	10 mW DC

disadvantage of EM radiation in this case is lower open circuit ac voltage at the terminals of the antenna compared to magnetic coupling or ultrasonic coupling. This requires additional voltage multiplier circuits to be able to turn on the electronics. Although ultrasonic coupling can provide higher power transfer efficiencies, especially for deep body implants [16], it requires a more sophisticated transducer setup compared to magnetic coupling. Considering the separation distance, the optimal solution is found to be employing magnetic coupling under the limitations given in Table 3.2 [20].

3.3 Magnetic Coupling

Upon deciding that magnetic coupling is the most suitable method for wireless power transfer in the target application, this section explains the fundamentals of this method by referring the literature. A theoretical background on the operation principle of inductive links and design considerations for the target application are also given. This section is concluded with the design and characterization of a 4-coil resonant inductive link.

Mechanism of magnetic coupling can be briefly explained using Faraday's law of induction as follows: a time-varying current on an inductor generates a time-varying magnetic field which then induces a time-varying voltage on another inductor which is present in the vicinity and if the second inductor is terminated with a load, circuit is completed and current passes through the load and wireless power transfer is therefore achieved. This is basically the operation principle of transformers which is an indispensable component of power distribution systems. In this book, magnetic coupling phenomena is exploited to create inductive links in order to transfer power from an external source to the intracranial neural implants. Inductive links can simply be formed using two inductors which are mutually coupled.

The most critical design goal for inductive links is to reach maximum power transfer efficiency under given restrictions dictated by the application such as operation frequency, separation distance, and implant coil size. In order to attack this optimization problem using a systematic approach, first, modeling of the inductive links has been studied to find out the maximum efficiency formulation. Currently, there are two analysis methods presented in the literature which are based on (1) coupled-mode theory [21] and (2) reflected load theory [22]. As indicated in [22], circuit designers are more familiar with the latter strategy since it allows a complete circuit representation of the entire wireless power transfer system. This section also employs reflected load theory to analyze the proposed system. Nevertheless, detailed formalism of coupled-mode theory has already been investigated in [21].

Inductances of the inductive links are commonly realized by two coils and they are placed in the vicinity of each other to create mutually coupled inductors. Several studies have been conducted to maximize the efficiency of the wireless power transfer systems incorporating 2 coils [23–27]. First of all, it has been shown that the two coils should be at resonance to maximize the efficiency as already proposed by

Nikola Tesla. Consequently, it can be claimed that first improvements in maximizing power transfer efficiency are realized by moving from nonresonant inductive links to resonant ones. Figure 3.2 illustrates the circuit representation of a 2-coil resonant inductive link.

L_1 and L_2 define the inductances created by two coils and R_1 and R_2 indicate their corresponding parasitic resistances. R_S and R_L denote the source resistance of the signal source and the load resistance, respectively. The inductors are brought to resonance at ω_0 with appropriate C_1 and C_2 capacitances while respecting the Eq. 3.1. In addition, quality factor is a characteristic parameter for resonant circuits and its expression is given in Eq. 3.2 for series resonant RLC circuits as given in Fig. 3.2.

$$\omega_0 = \frac{1}{\sqrt{LC}} \tag{3.1}$$

$$Q = \frac{1}{R}\sqrt{\frac{L}{C}} \tag{3.2}$$

Amount of coupling can be quantified by the coupling coefficient, k, which is represented by the mutual inductance, M, in circuit theory. Therefore, electrical modeling of this system involves mutually coupled inductors. Regarding the practical implementation of the mutually coupled inductors, there are numerous methods; however, not all of them fit with the requirements imposed for implantable systems. The limitation, mainly, arises in terms of the third dimension. Equation 3.3 introduces the relationship between mutual and individual inductances with the help of the coefficient k.

$$M = k\sqrt{L_1 L_2} \tag{3.3}$$

There are two major concerns in the design of an inductor for powering up an implant: (1) dimensions, and (2) efficiency. Here, efficiency, in fact, does not simply correspond to a dimensionless number; it includes all the information regarding supplied power, dissipated power on both inductors, and delivered power to the load. For instance, having a very high efficiency at a very high power level of dissipated

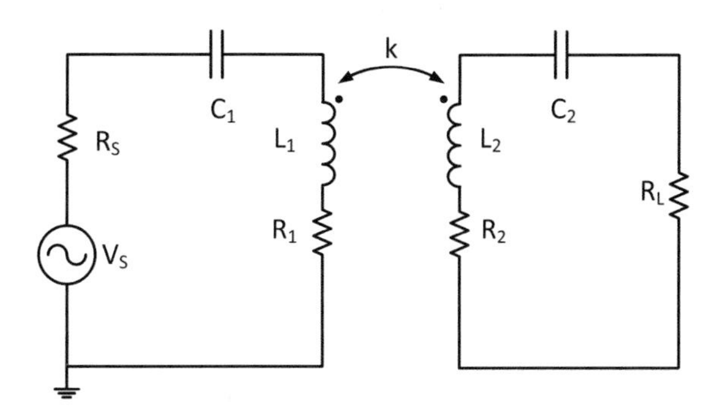

Fig. 3.2 Circuit representation of a 2-coil resonant inductive link

power on the implanted inductor may be deceiving since it may cause an intolerable temperature elevation in the surrounding tissues.

Maximum power transfer efficiency for a 2-coil resonant inductive link whose schematic is given in Fig. 3.2 can be expressed as:

$$\eta_{maximum} = \frac{U^2}{(1+\sqrt{1+U^2})^2} \tag{3.4}$$

where U can be defined as a function of the coupling coefficient and quality factors of each coil:

$$U = k\sqrt{Q_1 Q_2} \tag{3.5}$$

Therefore, for a given load condition, the coils should be designed accordingly or an impedance matching network should be employed as in [24]. As Eqs. 3.5 and 3.4 indicate, the efficiency depends on coupling coefficient (k) between coils and the quality factors (Q) of the coils. This leads us to geometric design optimization of spiral coils on PCB. Noting that the separation distance and implant size are generally dictated by the specific application, the remaining geometric parameters such as spiral shape, width, spacing, number of turns should be optimized [28]. Before going directly for geometric optimization, there is still room for improving the maximum efficiency equation by changing the conventional 2-coil resonant inductive link topology.

Recently, a modified version of resonant inductive links has been proposed for mid-range (a few meters) remote powering applications requiring tens of watts by [21]. The structure employing 4 coils instead of traditional 2-coil links has been adapted for implant powering applications [13]. The results exhibit a dramatic enhancement in the power transfer efficiency [22, 29, 30]. Figure 3.3 presents the lumped circuit models of 4-coil inductive links which are used to model such systems. As detailed in [29], 4-coil links compensate the low coupling coefficient and low Q-factor of source and load coils by using high-Q coils between them. Moreover, the efficiency does not change significantly with respect to source coils and more importantly load coil's quality factor. Additionally, they are less sensitive to distance variations compared to 2-coil links [29, 31]. Later, Kiani et al. [22] have shown that 3-coil resonant inductive links can provide almost the same power transfer efficiency by transforming any arbitrary load impedance to the optimal impedance needed at the input of the inductive link.

Maximum power transfer efficiency of a 4-coil resonant inductive link can be approximated as given Eq. 3.6 by neglecting the coupling coefficients k_{14}, k_{13}, and k_{24} [29]. This equation is induced by taking the equations of 2-coil resonant inductive links as basis.

$$\eta_{maximum} \approx \frac{(k_{12}^2 Q_1 Q_2)(k_{23}^2 Q_2 Q_3)(k_{34}^2 Q_3 Q_4)}{[(1+k_{12}^2 Q_1 Q_2)(1+k_{34}^2 Q_3 Q_4)+(k_{23}^2 Q_2 Q_3)][1+k_{23}^2 Q_2 Q_3+k_{34}^2 Q_3 Q_4]} \tag{3.6}$$

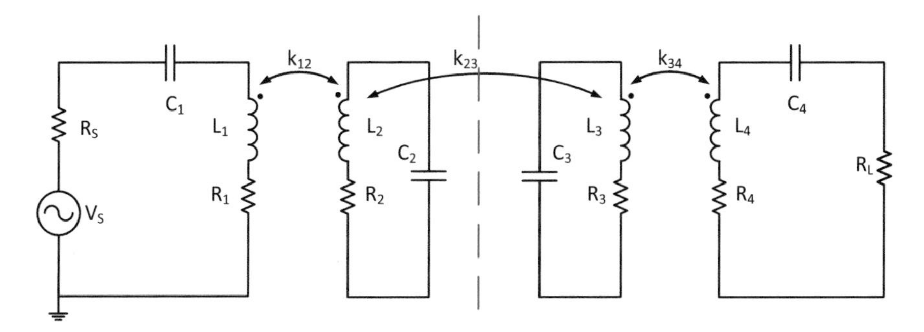

Fig. 3.3 Circuit representation of a 4-coil resonant inductive link

Equation 3.6 could be further simplified using certain assumptions. Firstly, we can assume that the coupling between the coils which are placed on the same PCB will be quite strong since they are coplanar. Therefore, it is safe to assume strong coupling between source and primary coils (k_{12}) and between secondary and load coils (k_{34}). Additionally, while designing the coils, it should be aimed to have high-quality-factor primary (Q_2) and secondary coils (Q_3). These first assumptions will yield the approximation in Eq. 3.7:

$$(1 + k_{12}^2 Q_1 Q_2)(1 + k_{34}^2 Q_3 Q_4) \approx (k_{12}^2 Q_1 Q_2)(k_{34}^2 Q_3 Q_4) \tag{3.7}$$

Moreover, it is also expected that coupling between primary and secondary coils (k_{23}) will exhibit a weak or moderate coupling, so under the condition that Eq. 3.8 holds;

$$(k_{12}^2 Q_1 Q_2)(k_{34}^2 Q_3 Q_4) \gg (k_{23}^2 Q_2 Q_3) \tag{3.8}$$

the following approximation for the first term of the denominator of Eq. 3.6 can be approximated as in Eq. 3.9:

$$[(1 + k_{12}^2 Q_1 Q_2)(1 + k_{34}^2 Q_3 Q_4) + (k_{23}^2 Q_2 Q_3)] \approx (k_{12}^2 Q_1 Q_2)(k_{34}^2 Q_3 Q_4) \tag{3.9}$$

For the second term of the denominator of Eq. 3.6, further simplification can be reached if Eq. 3.10 holds, which is also anticipated due to high-quality-factor primary (Q_2) and secondary coils (Q_3):

$$(1 + k_{23}^2 Q_2 Q_3) \gg (k_{34}^2 Q_3 Q_4) \tag{3.10}$$

Consequently, the second term of the denominator of Eq. 3.6 can be approximated as follows:

$$[1 + k_{23}^2 Q_2 Q_3 + k_{34}^2 Q_3 Q_4] \approx (1 + k_{23}^2 Q_2 Q_3) \tag{3.11}$$

Combining these two approximations and applying to Eq. 3.6 yields Eq. 3.12. As Eq. 3.12 shows, the maximum power transfer efficiency expression of 4-coil resonant inductive link can be approximately the same as the one of 2-coil resonant inductive link under certain conditions [29]. These conditions are having a high coupling coefficient between coils on the same side, e.g., L_1 and L_2, and having a high-quality factor for L_2 and L_3 coils. However, the advantage of using a 4-coil resonant inductive link is that it is almost independent of the quality factors of source and load coils which means that this efficiency can be achieved for a wider range of source and load resistances. Consequently, this creates a significant advantage over 2-coil structures at which maximum power transfer efficiency can be achieved under certain load conditions. Note that the neural recording system is prone to have different load currents (or impedances) during the operation since the number of recording sites and electrodes may vary.

$$\eta_{maximum} \approx \frac{k_{23}^2 Q_2 Q_3}{1 + k_{23}^2 Q_2 Q_3} \tag{3.12}$$

It is worth noting that the range of the load impedance, and hence load coil's quality factor, is not infinite. In order that Eq. 3.12 holds, the quality factor of the load coil (Q_4) is, in fact, limited by the requirements of the previous two approximations. Firstly, the approximation in Eq. 3.7 defines lower limit of Q_4 as given in Eq. 3.13:

$$(1 + k_{34}^2 Q_3 Q_4) \gg 1 \Rightarrow (k_{34}^2 Q_3 Q_4) \geq 10 \Rightarrow Q_4 \geq \frac{10}{k_{34}^2 Q_3} \tag{3.13}$$

Secondly, higher limit of Q_4 can be defined as in Eq. 3.15 to validate the approximation in Eq. 3.10:

$$(k_{23}^2 Q_2 Q_3) \gg 10 k_{34}^2 Q_3 Q_4 \Rightarrow Q_4 \leq \frac{k_{23}^2 Q_2}{10 k_{34}} \tag{3.14}$$

Finally, the boundaries of Q_4 can be defined by combining Eqs. 3.13 and 3.15 [29]:

$$\frac{10}{k_{34}^2 Q_3} \leq Q_4 \leq \frac{k_{23}^2 Q_2}{10 k_{34}} \tag{3.15}$$

As a result, in order to take the advantage of its higher power transfer efficiency and tolerance to a wide range of source and load impedances, a 4-coil resonant inductive link is implemented as the core of the wireless power transfer system. Initial implementations of the inductive links are based on winding of Litz wires as a helix. Litz wire is designed to reduce skin effect and proximity effect [29], so it can provide high-quality-factor coils. This 3D structure also enhances the uniformity of the magnetic field distribution which increases power transfer efficiency. However, later implementations employ spiral coils which are fabricated on printed circuit boards (PCB) since they occupy less volume and in fact they can be considered as

2D structures. Moreover, fabrication of spiral coils on flexible substrates enables conformal placement of large area coils on curved surfaces such as cortex of the brain. Alternatively, surface-mount inductors are also available; however, they are designed to control their flux inside a package to avoid interference with other circuits. Therefore, they are not suitable for this application. Although inductors with a higher thickness enables higher quality factors and coupling coefficient and more focused magnetic field lines, it is not feasible to put a solenoid in the brain implant, which is the targeted application. Therefore, the design is forced to be two-dimensional, which means that the third dimension has to be negligible compared to the other two dimensions. Moreover, creating the inductance on PCB favors the integration with the electronics which brings us the decision.

After deciding how to implement and the topology, the final step is to find the optimal geometry which will provide the maximum power transfer efficiency for a defined operation frequency. Operation frequency is selected as 8 MHz in order to minimize the tissue absorption [18]. For this geometric optimization, a code which has already been built in EPFL RFIC Research Group is employed [32]. Briefly, it runs a discrete optimization algorithm based on the divide-and-conquer search method to find the maximum value of the goal function [32]. Goal function in our case is the maximum power transfer efficiency defined in Eq. 3.12. Coupling coefficient and quality factor values are provided by the simulation tool FastHenry for any coil design. Note that certain parameters are not a part of optimization but are already fixed. Outer diameter for secondary coil is fixed by the dimension of the Burr hole and width and spacing of the metal lines are fixed by the manufacturing capability of PCB workshop.

Resulting coil dimensions are given in Table 3.3. Table 3.4 presents the electrical parameters of the designed rectangular spiral coils. Analytical calculations for the inductance of the coils using Wheeler equations [33] are found to be quite reliable following the measurements as shown in Table 3.4. In addition, quality factor of the coils and required external tuning capacitances to bring the four coils into resonance are presented. Designed inductive links have been manufactured using EPFL PCB workshop with respect to the fabrication parameters presented in Table 3.5. Finally, Fig. 3.4 introduces the images of the fabricated coils on FR4 PCB.

Before concluding the section, I would like to discuss some practical aspects of the decision of 4-coil resonant inductive links. 4-coil structures have a drawback

Table 3.3 Geometric parameters of the designed rectangular spiral coils [34]

	Inner diameter (mm)	Outer diameter (mm)	Width and spacing (mm)	Number of turns
Source coil (L_1)	13	24	0.5	6
Primary coil (L_2)	29	43	1	4
Secondary coil (L_3)	6.9	13.5	0.3	6
Load coil (L_4)	1.9	5.5	0.2	5

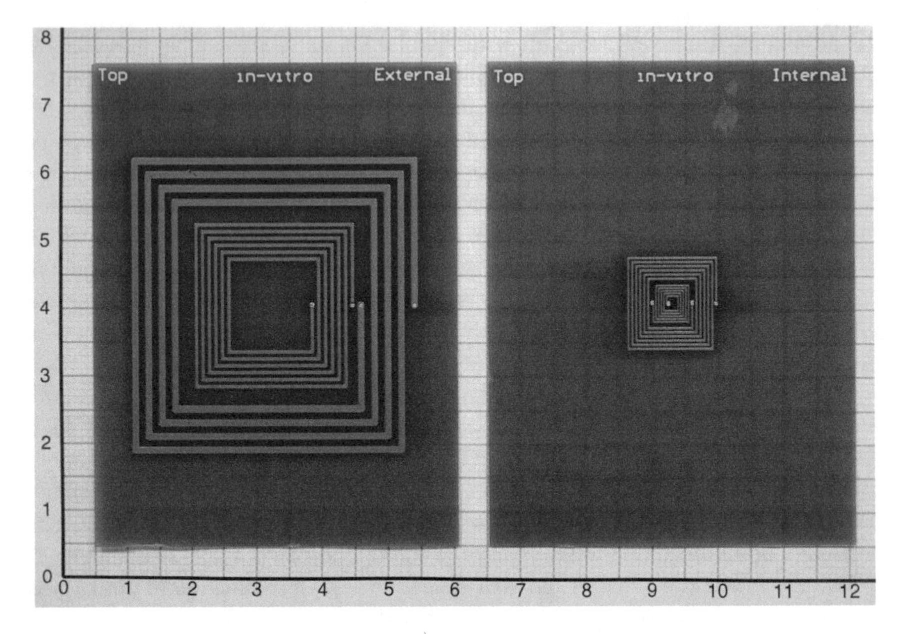

Fig. 3.4 Images of the fabricated coils on FR4 PCB: *left-outer* primary coil, *left-inner* source coil, *right-outer* secondary coil, and *right-inner* load coil [35]

compared to 2-coil resonant inductive links: It is, of course, more difficult to bring 4 coils into a certain resonance frequency than to bring 2 coils. The tuning procedure is, inherently, an iterative process due to the variations in the manufacturing. However, to reduce the number of iterations, I have used two approaches:

- using a parallel combination of two or three capacitances to reach the required tuning capacitance, which gives a higher degree of freedom than using a single capacitor and
- using a bank of capacitors which can be switched to reduce the time to find the required capacitance value to bring 4 coils into resonance.

Table 3.4 Electrical parameters of the designed rectangular spiral coils

	Inductance (μH) (analytical)	Inductance (μH) (measured)	Quality factor	External tuning capacitances (pF)
Source coil (L1)	1.07	1.10	48	330
Primary coil (L2)	1.10	1.14	64	326
Secondary coil (L3)	0.57	0.59	30	616
Load coil (L4)	0.12	0.14	21	2320

Table 3.5 Printed circuit board (PCB) manufacturing parameters for the designed 4-coil resonant inductive link

PCB manufacturation parameters	
Substrate type	FR4
Substrate thickness	0.8 mm
Metal (Cu) thickness	18 μm
Finishing	Sn

Note that the main difficulty arises when bringing the implant coils into resonance using minimum number of external components. On the other hand, it is comparably easier to bring the external coils into resonance since using two or three external capacitors does not violate size limitations for the external unit. I have designed a comparative experiment to observe whether it is worth the difficulty of tuning of 4-coil resonant inductive links compared to 2-coil resonant inductive links. Fixing the outer diameters of the coils, I have measured the power transfer efficiencies of 2-coil and 4-coil resonant inductive links at the same frequency and found them to be 37% for 2-coil and 55% for 4-coil resonant inductive links [13]. It can be concluded that the design difficulty can be justified with the obvious improvement in power transfer efficiency, which is quite critical due to temperature elevation regulations in the tissue. Moreover, a recent study comparing these two coil configurations indicates similar findings in terms of power transfer efficiency [22].

Another practical question is whether the maximum power transfer efficiency is really achievable, or not. It is practically quite difficult to bring all coil into the same resonance frequency unless a trimming system is utilized. Therefore, it can be claimed that finer-tuning could only result in a power transfer efficiency higher than 55%, but could not degrade this number since it is already slightly deviated from the optimal conditions due to practical limits.

3.4 Implantable Remote Powering Electronics

Terminating the load coil with a resistive element, an AC signal, and hence, power can be delivered to this load. However, a DC supply is required to power up the electronic circuits inside the implant. Therefore, this AC signal is first converted to a DC signal with ripples by means of a rectifier and then, this signal is converted to a stable DC signal using a regulator. This section gives a detailed explanation of these two circuits.

3.4.1 Rectifier

Induced AC voltage on the load coil is converted to a reliable DC voltage supply through a chain of circuits of which the rectifier constitutes the first module. Rectifier converts an AC signal to a DC signal on which reduced amplitude of source signal swings. A nonlinear component, commonly the diode, is used for rectification purposes. Common practice dictates that decreasing the turn-on voltage of the diode increases the efficiency. Therefore, a Schottky diode, which possesses a lower turn-on voltage, is practical to employ for discrete circuits. However, the fabrication of Schottky diodes is not favorable for standard CMOS processes. Therefore, the Schottky diode is generally replaced with a diode-connected transistor despite the threshold voltage drop-out in passive rectifiers or with a pass transistor working as a switch to lower the drop-out voltage in active rectifiers for the sake of integration.

Active rectifiers are preferred if the increase in efficiency with respect to the passive rectifier compensates the dissipated power in decision mechanisms utilized in active rectifiers. It can be concluded that as the power entering the rectifier increases, it is more advantageous to use an active rectifier since the power budget on the decision mechanism is kept almost constant even if the input power increases [36].

Overall efficiency of the wireless power transmission link depends on the efficiency of each component of the link. Since power conversion efficiency of the rectifier varies under different operation conditions, it is a critical issue to be considered in the predesign phase. More explicitly, it is a quite common observation that efficiency increases up to a certain extent as the input power increases for the same rectifier. It occurs in the circuits for which the power loss does not increase in the same rate as the input power increases. Furthermore, keeping the input power constant, variation of the input voltage level may result in different efficiencies under different loading conditions if the rectifier has a constant voltage drop between its input and output terminals, as in the case of a diode. Priority of the mentioned parameters varies according to the application. Another important design parameter is the input voltage level which enables proper operation. However, design flexibility on this parameter can be obtained via adjusting the voltage conversion ratio of mutually coupled inductances which is related with their inductance values.

Consequently, in this application, the power conversion efficiency of the rectifier is monitored by keeping the output DC voltage of the rectifier constant under varying loads since

- efficiency of the following regulator is therefore preserved which is critical in terms of system-level design,
- power feedback loop compares this point with a reference voltage to sustain proper and efficient operation, and
- it simulates real operation scenarios in which active number of channels and chips of the recording unit vary, and thus affecting the required power.

In the scope of this work, an active half-wave rectifier has been designed by taking the structure presented in [37] as an initial point. Half-wave rectifier provides a higher efficiency compared to its full-wave version since the drop-out voltage is lower. The rectifier consists of three major modules: (a) switching element with dynamic bulk biasing, (b) comparison, and (c) timing and decision. Figure 3.5 presents the interconnections between these three modules. Comparison module compares input and output voltages continuously and then, timing and decision module controls

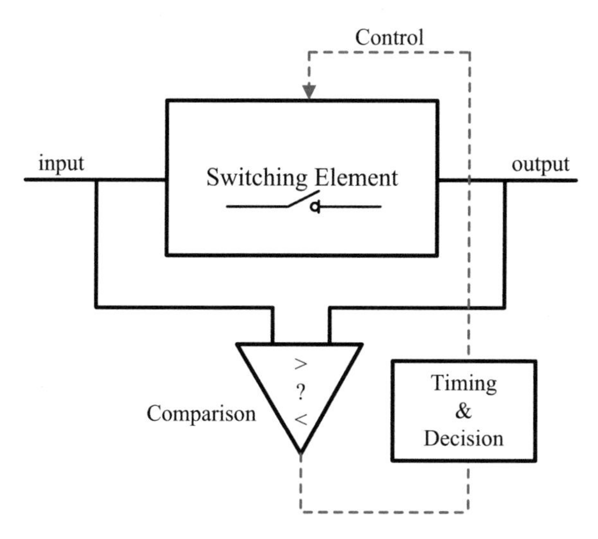

Fig. 3.5 Operation scheme of the active rectifier

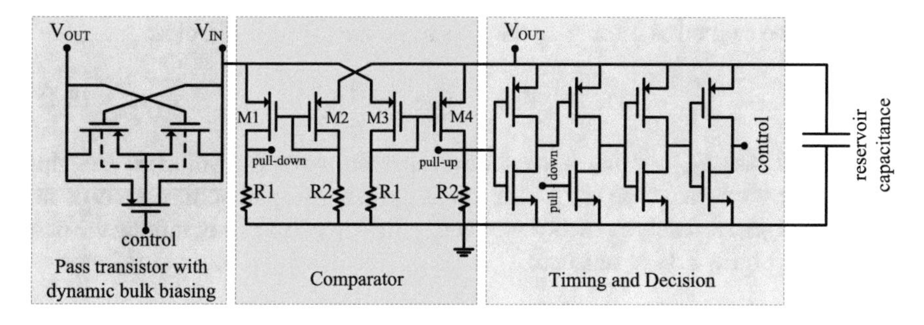

Fig. 3.6 Schematic of the active half-wave rectifier [34]

the duration how long the switch will be turned on and off to charge the reservoir capacitance.

A PMOS transistor is employed as the switching element and its bulk bias is provided via dynamic bulk biasing method [38] as depicted in Fig. 3.6. Using a PMOS transistor prevents start-up problems since it starts to conduct when the gate potential is zero and also, it eliminates the necessity of possessing a twin-well process for dynamic bulk biasing. Dynamic bulk biasing is necessary to prevent latch-up, especially in the start-up. During start-up, input and output voltages exceed each other sequentially. Thus, bulk of the PMOS transistor requires a dynamic connection to the highest potential at that instant. Although it is not so critical in steady-state, dynamic bulk biasing is not turned off since it prevents a leakage from the input to the bulk by connecting them together. The PMOS pass transistor is switched on when the input voltage is higher than the output voltage so that the reservoir capacitance is charged. It is switched off otherwise to prevent leakage from the output to the input.

Comparator decides whether to pull-up or pull-down the control voltage, which biases the gate of the PMOS pass transistor, according to the instantaneous input and output voltages [39]. For instance, control voltage should be pulled down to the lowest potential available, which is ground in this case, to allow the pass transistor to conduct when the input voltage is higher than the output voltage. Therefore, the reservoir capacitance could be charged. For the other case, control voltage should be pulled up to the highest potential available, which is the rectifier output voltage in this case, to turn off the pass transistor to prevent backward leakage current.

Operation principle of the half circuit of the comparator is as follows: The current passing through diode-connected M_2 is quite low due to high resistance at the drain. Thus,

$$V_{G2} \approx V_{out} - |V_{th2}| \tag{3.16}$$

M_1 turns on and hence, pull-down voltage is generated if Eq. 3.17 holds:

$$V_{in} - V_{G2} \ > \ |V_{th1}| \tag{3.17}$$

Combination of these two inequalities yields a pull-down operation, which can eventually be defined as turning on the pass transistor, if Eq. 3.18 holds:

$$V_{in} \ > \ V_{out} - |V_{th2}| \ + \ |V_{th1}| \tag{3.18}$$

Provided that M_1 and M_2 have almost the same threshold voltages, charging of the reservoir capacitance starts immediately when the input voltage exceeds the output voltage. The analysis of the other half of the circuit results in turning the pass transistor off if Eq. 3.19 is satisfied:

$$V_{out} \ > \ V_{in} - |V_{th3}| \ + \ |V_{th4}| \tag{3.19}$$

Provided that M_3 and M_4 have almost the same threshold voltages, reverse current is minimized since the switch will be turned off if input voltage goes below the output voltage.

Following the comparator, a multiplexer with an output buffer finalizes the circuit and closes the loop by creating the control signal which is driving the pass transistor. Proper operation of the rectifier requires that the pass transistor should be switched on only when the input voltage is higher than the output voltage so that the reservoir capacitance employed at the output can be charged and should be switched off otherwise to prevent leakage from the capacitor to the input. Generated pull-down and pull-up voltages in the comparator are both high level regardless of the condition. Therefore, pull-up voltage is inverted and supplied to a PMOS while pull-down voltage is directly connected to a NMOS as presented in Fig. 3.6. Therefore control voltage is either connected to the rectifier output which turns the pass transistor into a diode-connected transistor and prevents leakage to the input or connected to the ground which turns the pass transistor on completely to charge up the reservoir capacitance. Its performance can be evaluated with its speed and timing of switching.

Switching speed is increased by adding a buffer to charge the gate capacitance of the pass transistor faster at the expense of the dynamic power consumption of the buffer. As a result, through post-layout simulations, rise time and fall time of the control signal driving the pass transistor have been calculated as 3.4 and 1.7 ns, respectively. Due to the nature of the rectification process, the switch has to once turn on and once turn off in a half cycle of the input signal. Since the operation frequency of the wireless power transfer system is 8.5 MHz, the half-period corresponds to 58.8 ns. Since the sum of rise and fall times (5.1 ns) are smaller than one-tenth of the half cycle of the input signal, the switching capabilities of the pass transistor are acceptable as sufficiently fast.

However, the accuracy of timing mainly depends on the comparator. Here, comparator provides the select signals of the multiplexer: pull-up to switch off and pull-down to switch on. Operation principle of the comparator relies on biasing the diode-connected transistors with very low current thanks to the high resistances at the drain end such that voltage headroom on those transistors can be approximated to their thresholds. With proper layout techniques, thresholds of all transistors in the comparator can be made almost the same, and then, pull-down and pull-up voltages follow the same trend with input and output voltages, respectively. Consequently, matching the layout of these four transistors plays a significant role in the accuracy of timing.

In order to analyze the performance of the two blocks, namely comparator and timing and decision, in terms of creating the control signal, which will bias the gate of the PMOS pass transistor, a post-layout transient simulation has been run and resulting regulator input and output waveforms, as well as the control voltage waveform (see Fig. 3.6) are presented in Fig. 3.7. As shown in Fig. 3.7, control signal has a latency of 3 and 5 ns for turning on and off, respectively. Figure 3.7 presents an additional information about the case when V_{in} becomes equal to V_{out}, both for an increasing and decreasing V_{in}. Note that, the comparator output, or more precisely the control voltage, tends to preserve its current state when V_{in} becomes equal to V_{out}. More explicitly, if this scenario occurs when V_{in} has an increasing trend, the control voltage stays high, by leaving the pass transistor off till V_{in} exceeds V_{out} just by the amount of the deadzone. As every comparator, this one has also a deadzone where comparator gain is not sufficient to saturate the amplifier to one of the supply voltages. On the other hand, if this scenario occurs while V_{in} has a decreasing trend, then control signal stays low, keeping the pass transistor on till the passage through the deadzone is over.

Following the schematic-level design of the circuit, its layout has been drawn and following the parasitic extractions, post-layout simulations to depict the power conversion efficiency (PCE) ($P_{OUT,DC}$ / $P_{IN,AC}$) with respect to output power have been performed for different operation frequencies ranging from 4 to 12 MHz with 2 MHz steps. As Fig. 3.8 shows, PCE becomes maximum around 10–12 mW output power. It is worth noting that a series resistance of 10 Ω is added to the input of the rectifier in order to mimic practical test conditions. Furthermore, the parasitic inductance and capacitances of the bonding wires are added to the simulation environment to achieve more realistic results. As mentioned before, output DC voltage

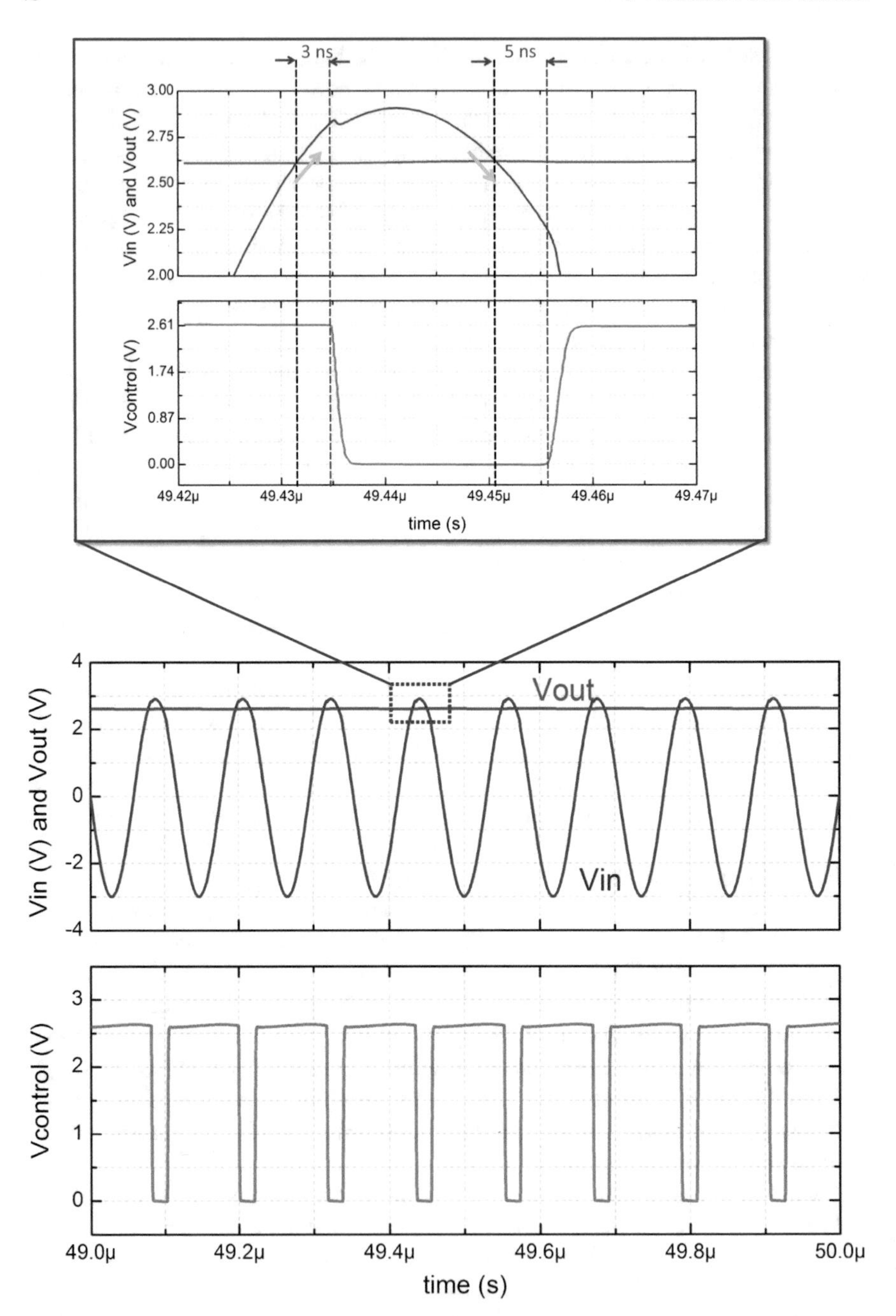

Fig. 3.7 Waveforms explaining the latency of the control voltage generation which turns on and off the PMOS pass transistor as well as the comparator deadzone; for both cases of V_{in} becomes equal to V_{out}, while both increasing and decreasing

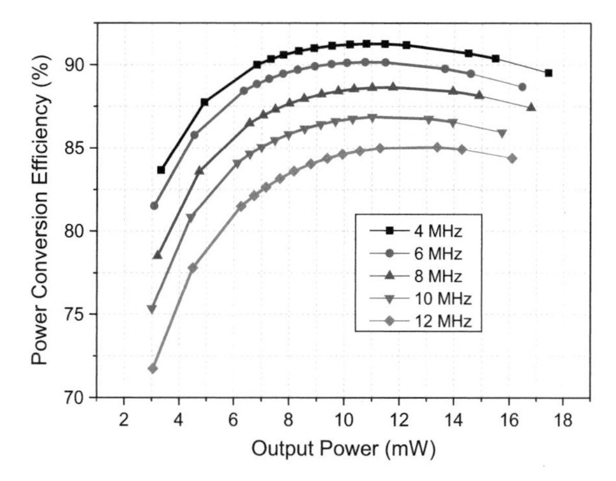

Fig. 3.8 Power conversion efficiency of the rectifier with respect to output power for different operation frequencies

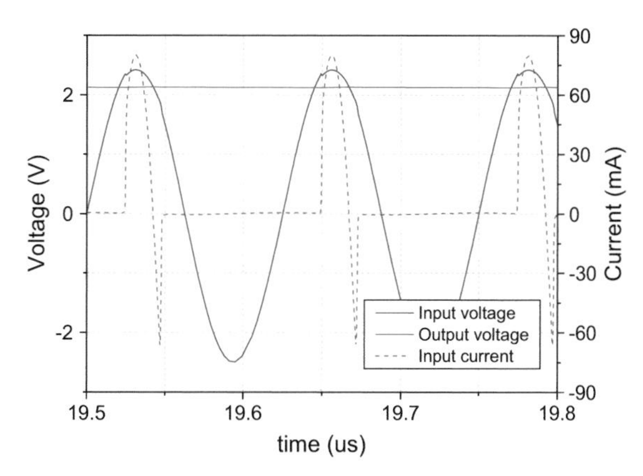

Fig. 3.9 Waveforms of input voltage, input current, and output voltage at steady state

is kept constant (at 2 V for this case) while performing the post-layout simulations to observe the variation of PCE with respect to the output power.

Figure 3.9 shows the steady-state time-domain signals for input voltage (V_{in}), input current (I_{in}), and output voltage (V_{out}) under the nominal load. The nominal output power is 10 mW while the output DC voltage is 2 V for 12 MHz excitation. Note that there is a certain amount of backward leakage current which in fact degrades the power conversion efficiency. This backward leakage current occurs due to the deadzone in the comparator and the delay of the circuits. The duration between decision and execution is equal to the duration of this backward leakage current. It can be avoided by deciding to turn off the pass transistor before the equality of input voltage to output voltage happens. At first glance, it seems it is against causality principle; however, it can be realized, for instance, by comparing $0.9 \times V_{in}$ with V_{out} instead of a direct comparison. The voltage division ratio could be determined from the post-layout simulations in order to compensate the delay between decision and

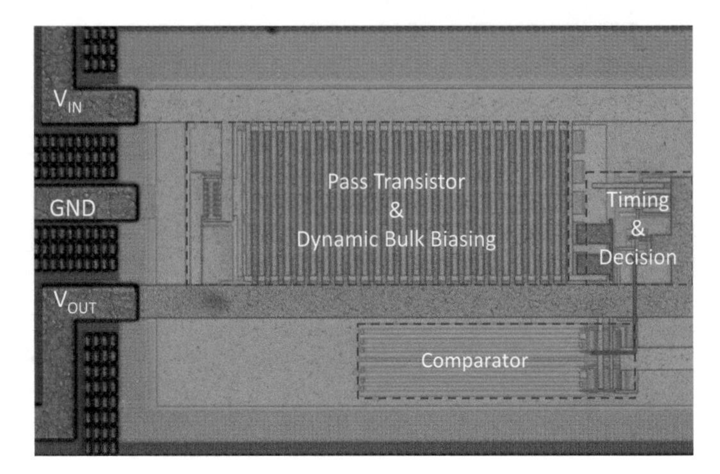

Fig. 3.10 Micrograph of the active half-wave rectifier which is fabricated with UMC 180 nm technology

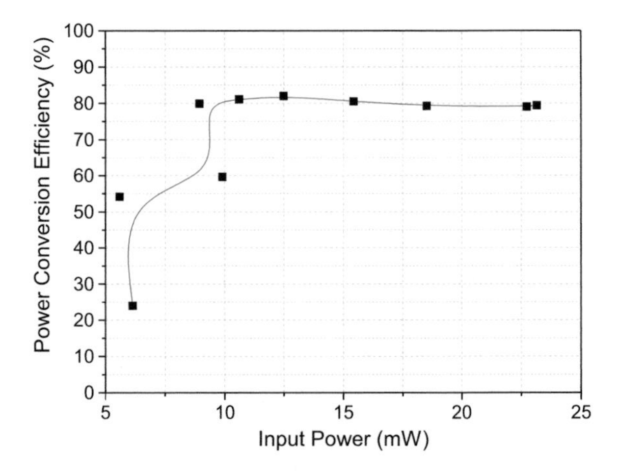

Fig. 3.11 Measurement results of the active half-wave rectifier showing the power conversion efficiency with respect to input power while output voltage is kept constant at 2 V

execution. However, it requires a proper voltage division, possibly with an operational amplifier which requires a dc supply.

The UMC 180 nm manufacturing process was used to design the active half-wave rectifier circuit. Figure 3.10 shows the micrograph of the fabricated rectifier with annotations of three major blocks. It occupies an area of 110 μm $\times$ 250 μm on silicon.

Figure 3.11 shows the measurement results obtained to characterize the power conversion efficiency of the rectifier with respect to input power while delivering output power ranging from 1 to 20 mW; output voltage is kept constant at 2 V. In this case, AC excitation frequency is set to 8 MHz.

Table 3.6 Comparison of the designed half-wave active rectifier with similar works in the literature

Ref.	Process technology (nm)	Operation frequency (MHz)	Input voltage (Vp)	Reservoir capacitance (nF)	Output voltage (V)	Output power (mW)	Simulated power conversion efficiency (%)	Measured power conversion efficiency (%)
[40]	180	13.56	N/A	5800	1.2	112.5	93	N/A
[41]	350	13.56	1.5–3.5	0.2	1.2–3.22	5.76	65–89	N/A
[42]	350	1.5	1.2–2.4	1000	1.13–2.28	43.3	82–87	N/A
[43]	500	13.56	3.3–5	10000	2.5–3.9	30.42	71–84.5	68–80.2
[44]	180	13.56	0.9–2	10000	0.45–1.78	3.2	N/A	60–81.9
[45]	180	10	0.8–2.7	0.2	0.2–2	2	60–86	37–80
[46]	350	13.56	1.5–4	1.5	1.19–3.52	24.8	84.2–90.7	82.2–90.1
[44]	180	13.56	1.5	10000	1.33	1.7	N/A	81
This work	180	8.5	2.5	100	2.1	13	88	82

Finally, Table 3.6 summarizes and compares the performance of the implemented rectifier with the state-of-the-art rectifier designs mainly targeting wireless power transfer systems.

3.4.2 *Voltage Regulator*

The output of the rectifier is composed of a DC voltage with a ripple at the operation frequency. In order to eliminate this ripple and provide a DC voltage independent of the input voltage and of the load resistance, a regulator has to be employed. A reliable DC supply is quite critical for proper operation of analog-to-digital converters (ADC) in the sensor unit. As explained in [31, 34], the power efficiency of the blocks on the energy transfer link is the most critical design goal; therefore, a low drop-out voltage (LDO) regulator is designed to provide a reliable DC supply.

The designed regulator aims to provide a 1.8 V regulated output DC voltage. Figure 3.13 presents the schematic of the LDO regulator biased with a cascoded bootstrapped current source. Since the application is intrinsically intolerant to large temperature variations [47], supply-independent biasing is preferred instead of a voltage reference circuit such as bandgap reference. Reference current is mirrored to bias the operational transconductance amplifier (OTA) which works as an error amplifier. OTA compares the reference voltage with the sampled voltage of the output and drives the PMOS transistor to sustain the output voltage constant at 1.8 V. As a

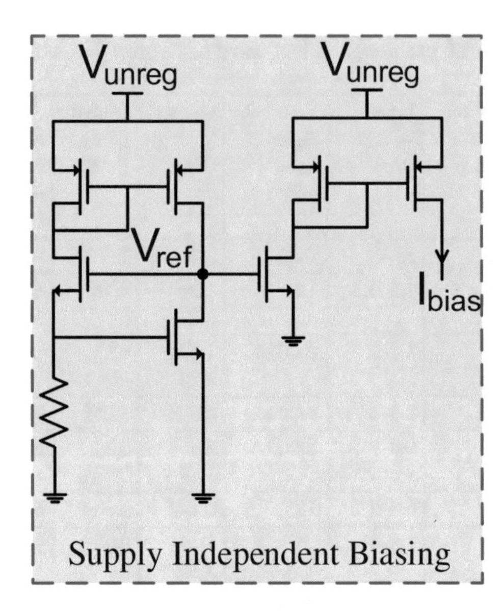

Fig. 3.12 Schematic of the V_{GS}-referenced supply-independent voltage reference circuit

result, the unregulated voltage is purified from its ripple content and a constant DC voltage is obtained.

Firstly, a V_{GS}-reference supply-independent current reference circuit is employed as a voltage reference as proposed in [48]. Figure 3.12 depicts the first version of the voltage reference circuit.

The decision of using a bootstrapped current source relies on the assumption that the rectifier output voltage will not vary significantly during the operation. However, the channel length modulation of the upper current mirror transistors prevented us from reaching the desired power supply rejection ratio (PSRR) which is critical for ADC operation. In the second design version, the circuit is improved by cascoding the upper mirror transistors. Figure 3.13 shows the modified version of the voltage reference circuit with the start-up circuit.

Comparison of the reference voltage and the sampled regulator output voltage is performed by an operational transconductance amplifier (OTA). OTA switches on or off the pass transistor according to the comparison in order to keep the output voltage constant. Overall PSRR of the regulator is dominated either by the PSRR of the voltage reference circuit or the gain of the OTA. Consequently, the performance of the OTA is critical to avoid degrading the performance of the voltage reference. A single-ended differential amplifier has been designed as the error amplifier as shown in Fig. 3.13. Note that noise from the power supply is a common-mode signal, and single-stage topologies offer better common-mode rejection ratio (CMRR) than multi-stage versions. Lower power consumption and not requiring an additional compensation capacitor can be listed among the advantages which led me into this decision.

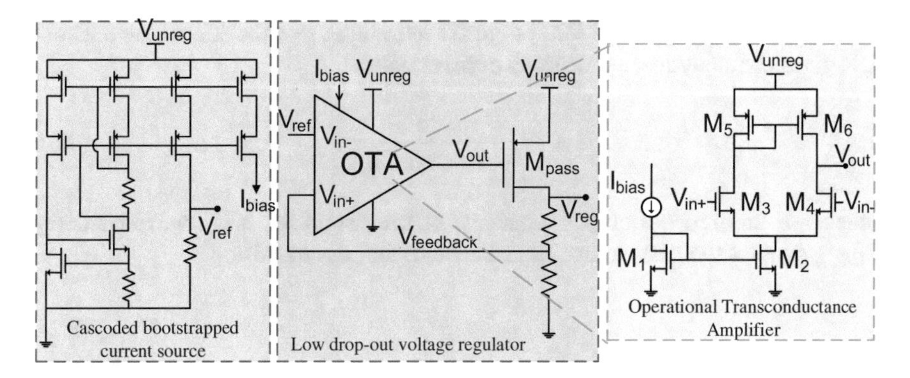

Fig. 3.13 Schematic of the low drop-out voltage regulator with the cascoded bootstrapped current source [34]

Design of the OTA is performed by exploiting a g_m/I_D-based approach which enables design in all operations regions of the MOSFET [49, 50]. In addition, EKV model [51], which enables an extensive modeling of the transistor from weak inversion to strong inversion, and therefore which enables low-voltage low-power analog circuit design, is employed. g_m/I_D method relates the transconductance (g_m) over drain current (I_D) ratio to a factor called inversion coefficient (IC) which is a function of normalized drain current ($I_D/(w/L)$). More explicitly, two main equations which are used in transistor sizing are as follows:

$$\frac{g_m}{I_D} = \frac{1}{nU_T} \cdot \frac{(1 - e^{(-\sqrt{IC})})}{\sqrt{IC}} \tag{3.20}$$

$$IC = \frac{I_D}{2n\mu C_{ox}\frac{W}{L}} U_T^2 \tag{3.21}$$

where C_{ox} defines the gate oxide capacitance per unit area, U_T is the thermodynamic voltage (26 mV at room temperature), and n is the slope factor.

First g_m is determined from design specifications. Then, a moderate channel length, L, is chosen both to have a considerable speed and to match, especially the input pairs well to minimize input offset. In order to increase power conversion efficiency and reduce quiescent current, I_D is chosen as low as possible. Finally, widths of the transistors, w, are determined using the previous parameters.

Stability analysis of the regulator can be performed by locating poles and zeroes of the system function. First pole of the circuit is defined by the external components at the output, i.e., load capacitance (C_{load}) and resistance (R_{load}). Equation 3.22 presents the first pole of the circuit:

$$\omega_{P_1} = \frac{1}{R_{load}\ C_{load}} \tag{3.22}$$

Second and third poles of the system are introduced by OTA. Equations 3.23 and 3.24 gives the analytical expressions of these poles.

$$\omega_{P_2} \approx \frac{1}{(ro_4||ro_6)\ C_{GS_{pass}}} \tag{3.23}$$

where ro_4 and ro_6 are output resistance of transistors M_4 and M_6, respectively. $C_{GS_{pass}}$ denotes the gate–source capacitance of the pass transistor.

$$\omega_{P_3} \approx \frac{g_{m_5}}{3\ C_{GS_5}} \tag{3.24}$$

where g_{m_5} is the transconductance of the transistor M_5 and C_{GS_5} denotes the gate–source capacitance of the same transistor.

In order to ensure stability, these two parasitic poles, which are introduced by the OTA, should be compensated by means of a zero. The only zero of the system is derived from the load capacitance equivalent series resistance (ESR) and the load capacitance (C_{load}) itself. Equation 3.25 expresses the angular frequency of this zero analytically.

$$z_1 = \frac{1}{R_{ESR}\ C_{load}} \tag{3.25}$$

Following the schematic-level design of the low drop-out voltage regulator series of post-layout simulations has been conducted to characterize the system and to analyze its stability. In terms of stability, one way is to plot the Bode plots of the system, while the more reliable method is to verify it through long transient simulations. Firstly, a post-layout simulation has been designed to see the output of the regulator at maximum current demand, 5 mA, for a duration of 1 ms whose results have been depicted in Fig. 3.14. Secondly, same simulation has been conducted with the minimum current demand, 1 mA, for a duration of 1 ms. Figure 3.15 shows the waveforms obtained for regulator input and output voltage, as well as the output current demand. Note that the input voltage is defined as a 20 mV AC signal at the operation frequency (8.5 MHz) on top of 2.1 V DC voltage. Current demands set by considering all sensor units are active or all passive for maximum and minimum current demand cases, respectively.

Following the stability analysis, three additional characterization testbenches are designed to observe (1) the performance of the regulator for varying DC input voltage conditions (line regulation), (2) varying output current demand (load regulation), and (3) a transient load response which emulates a scenario in which input voltage and output current demand change simultaneously. Figure 3.16 shows the post-layout simulation results for variation of regulator output voltage with respect to input voltage ranging from 2.1 to 2.9 V. In Fig. 3.17, the post-layout simulation results for variation of regulator output voltage with respect to output load current ranging from 1 mA to 5 mA has been presented.

Finally, Fig. 3.18 presents the transient load simulation and the voltage regulator response to these simultaneous variations.

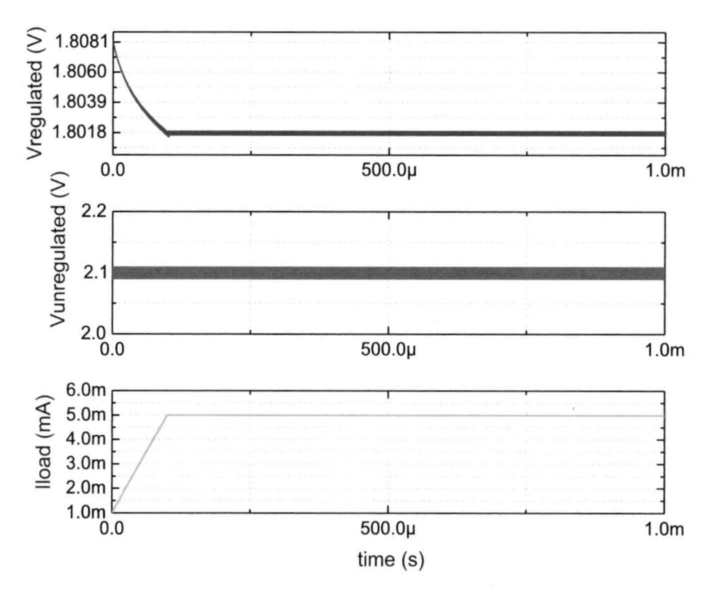

Fig. 3.14 Post-layout transient simulation results to show the stability of the regulator for maximum current case while input DC voltage is 2.1 V

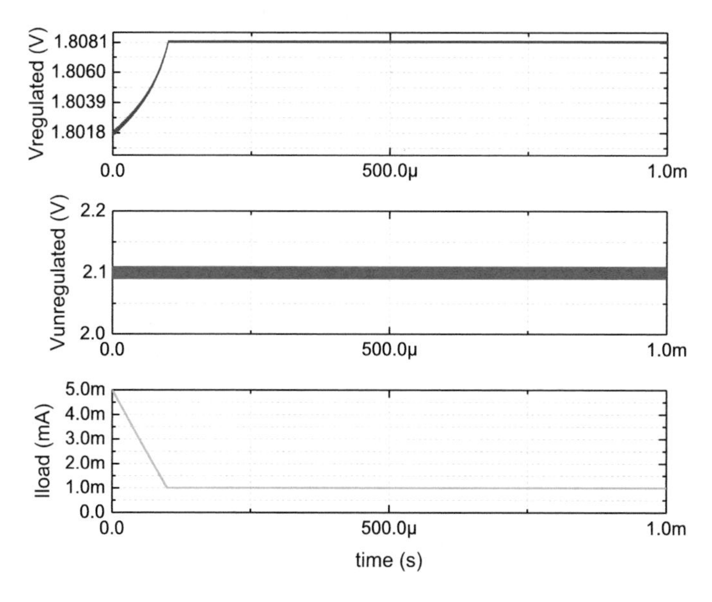

Fig. 3.15 Post-layout transient simulation results to show the stability of the regulator for minimum current case while input DC voltage is 2.1 V

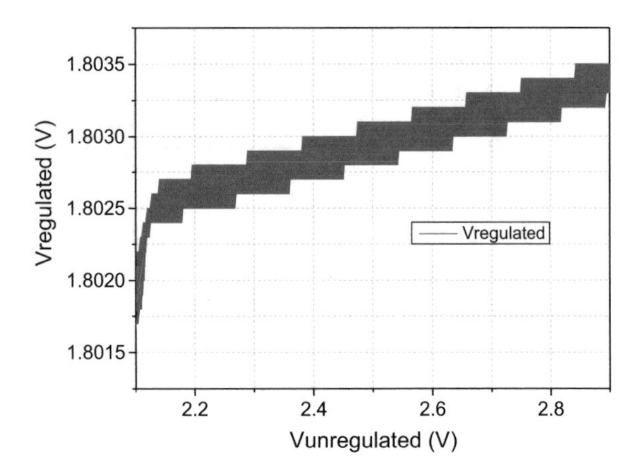

Fig. 3.16 Post-layout simulation results for variation of regulator output voltage with respect to input voltage ranging from 2.1 V to 2.9 V

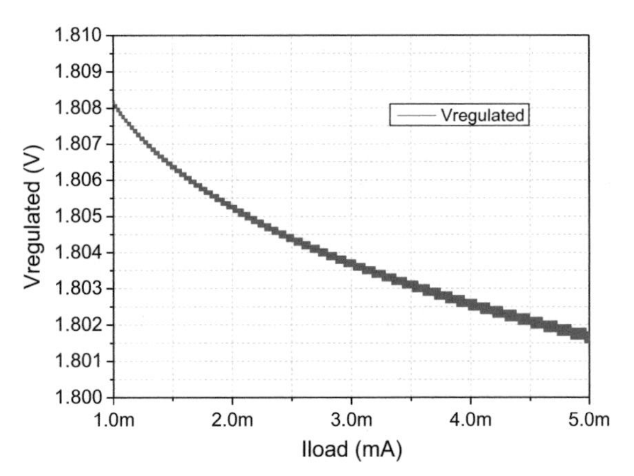

Fig. 3.17 Post-layout simulation results for variation of regulator output voltage with respect to output load current ranging from 1 mA to 5 mA

Designed low drop-out voltage regulator has been manufactured using UMC 180 nm MM/RF process technology. Figure 3.19 shows the micrograph of the fabricated regulator with annotations of the major building blocks. It occupies an area of 125 μm $\times$ 200 μm on silicon.

Aforementioned, efficiency of the regulator is critical since it is located in the implant side. Efficiency of this regulator can be estimated as (V_{out} / V_{in}) by neglecting the power consumption of the OTA and sampling resistors (<100 μW) which is reasonable for 10 mW output power. As a result, it is expected to reach 82% efficiency under our operation conditions while DC level of the input voltage is 2.2 V. Figure 3.20 presents the simulated power conversion efficiency of the regulator with respect to the output power for different DC levels at the input. Power conversion efficiency of the regulator for 10 mW output power is measured to be 78% while delivering 10 mW to the output load.

Moreover, power supply rejection ratio (PSRR) of the regulator is another critical parameter to be evaluated. It has been measured with respect to the frequency under

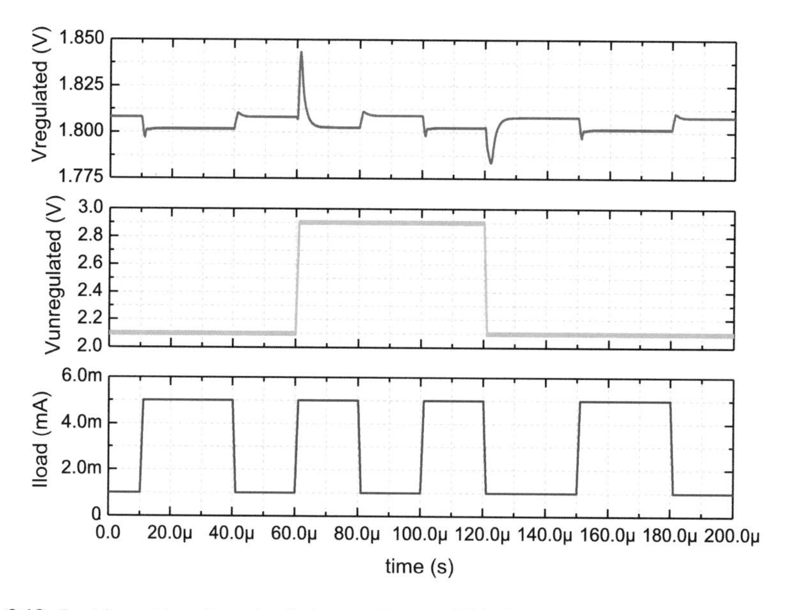

Fig. 3.18 Post-layout transient simulation results to exhibit the response of the regulator for fast changing current demand and input voltage variations

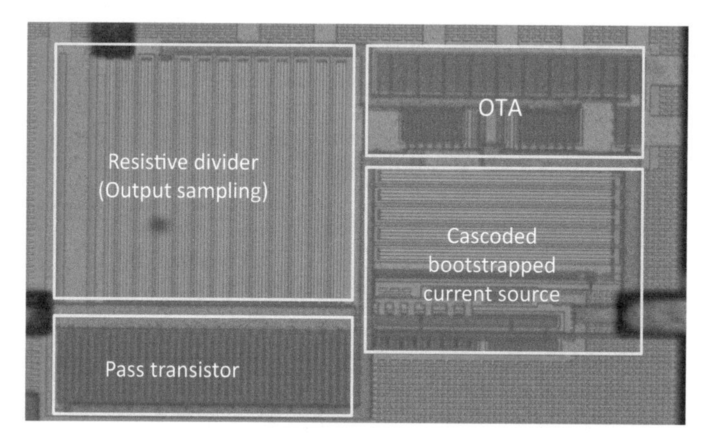

Fig. 3.19 Micrograph of the low drop-out voltage regulator fabricated with UMC 180 nm technology

2.2 V input DC bias using Texas Instruments THS3120EVM board. The output load of the regulator is set to 330 Ω and 100 nF. Figure 3.21 shows the power supply rejection ratio of the regulator. It provides a suppression of 50 dB around DC and 60 dB around operation frequency of 8 MHz. Moreover, load regulation is calculated to be 0.15% for an output power range from 1 mW to 10 mW.

Considering that target application is going to employ an analog-to-digital converter (ADC) to digitize the acquired neural signals, it is reasonable to evaluate the practical performance of the integrated regulator by driving an ADC. Therefore, a

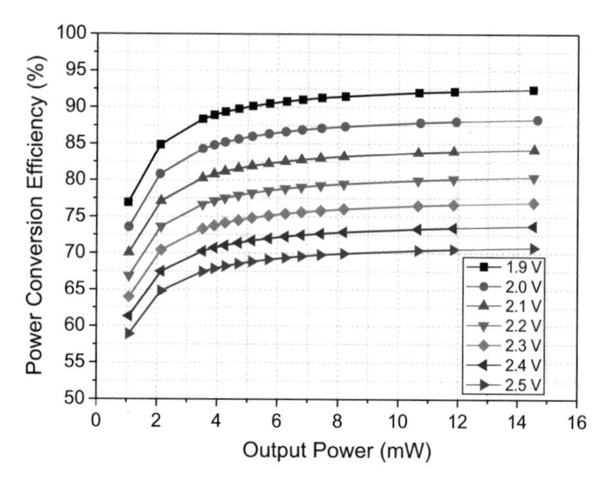

Fig. 3.20 Simulated power conversion efficiency of the regulator with respect to the output power for different DC levels at the input

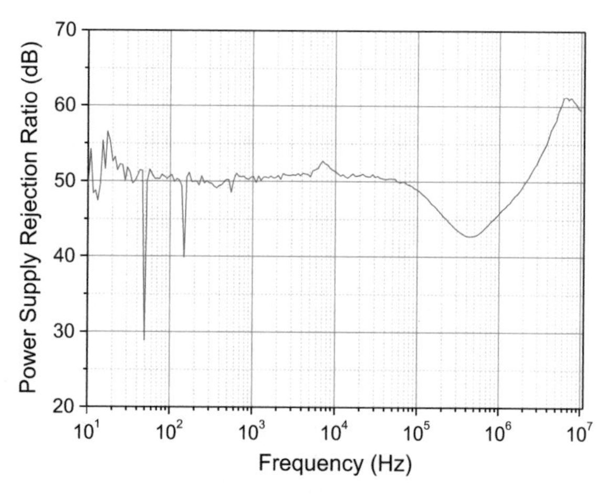

Fig. 3.21 Measured power supply rejection ratio (PSRR) of the regulator with respect to the frequency when 2 V DC is applied at the input

commercial ADC (AD7466) is supplied by the designed regulator to observe how many bits stay stable for a given input voltage while the voltage level of the signal at the input of the regulator is changing. It is, in fact, another measure of line regulation or PSRR. Therefore, rectifier output voltage has been increased by 1 V starting from the most power efficient case. This voltage difference has been chosen by considering the voltage variations due to possible misalignments during operation and to enable modulations with high modulation index. Clock signal for the ADC is generated using the recovered clock and the chip select is generated externally. Consequently, we have observed that 8 bits out of 12 always stay stable if the regulator input (or the rectifier output) increases by 1 V. The measurements were repeated for different analog voltage levels, and it was observed that in some cases, where the analog voltage corresponds to the center of gravity of a particular binary number, 9 bits stay stable. Consequently, we conclude that the regulator is sufficient to drive an 8-bit ADC while least significant bit (LSB) is calculated to be 7 mV.

Table 3.7 Comparison of the designed low drop-out voltage regulator with similar works in the literature

	Process technology (nm)	Unregulated input voltage (V)	Regulated output voltage (V)	Load current (mA) (max)	Quiescent current (μA)	Load capacitance (pF)	PSRR (dB)	Line regulation (mV/V)	Load regulation (mV/mA)	Power conversion efficiency (%)
[52]	90	0.75	0.5	100	8	50	44 @ 1 kHz	3.78	0.1	66
[53]	90	1.2	0.9	100	6000	600	N/A	N/A	1.8	75
[54]	350	3	2.8	50	65	100	57 @ 1kHz	23	0.56	93
[55]	350	1.8	1.6	100	20	100	40 @ 10 kHz	0.0574	0.109	88
[56]	350	1.4	1.2	100	43	1000	N/A	N/A	0.4	85
[57]	350	0.9	0.5	50	0.103	0	N/A	21.76	0.324	55
[58]	350	1.65	1.5	100	27	100	39.5 @ 10 kHz	1.046	0.0752	90
[59]	65	1.2	1	100	82.4	100	58 @ 10 kHz	4.7	0.3	83
This work	180	2.2	1.8	5	100	100000	50 @ 10 kHz	1	1.6	82

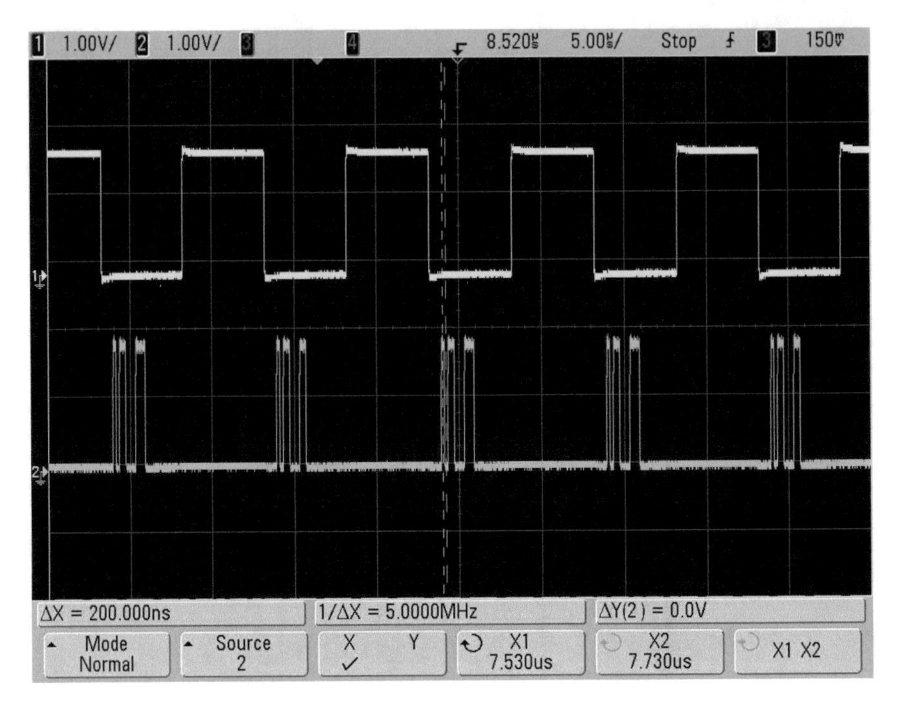

Fig. 3.22 ADC output of an incoming signal at 1234 mV (1011 0001 1XXX)

Finally, Table 3.7 summarizes and compares the performance of the implemented low drop-out voltage regulator with the state-of-the-art regulator designs mainly targeting wireless power transfer systems (Fig. 3.22).

3.5 Summary

This chapter explains how wireless power transfer is realized for the targeted intracranial neural recording system. Firstly, the motivation for using the wireless power transfer technique versus ambient energy harvesters and medical-grade batteries is justified in accordance with the constraints on the specifications discussed in Chap. 2. Next, it is justified that magnetic coupling is the most appropriate solution to remotely power an intracranial implant taking tissue absorption concerns and separation distance into account. Realization of the wireless power transfer system using 4-coil resonant inductive link and a power supply generation unit in the implant, consisting an active half-wave rectifier and a low drop-out voltage regulator, is explained in detail. Additionally, individual performance characterizations of the circuits have been introduced. Contents of this chapter have been published in [13, 31, 35, 60].

References

1. Massachusetts General Hospital, Brain electrodes for epilepsy surgery, http://www.massgeneral.org/neurology/assets/ResearchLabs/cash_MG51brainElectrodesFiltered.jpg
2. K. Murakawa, M. Kobayashi, O. Nakamura, S. Kawata, A wireless near-infrared energy system for medical implants. IEEE Eng. Med. Biol. Mag. **18**(6), 70–72 (1999)
3. K. Goto, T. Nakagawa, O. Nakamura, S. Kawata, An implantable power supply with an optically rechargeable lithium battery. IEEE Trans. Biomed. Eng. **48**(7), 830–833 (2001)
4. S.-N. Suzuki, T. Katane, H. Saotome, O. Saito, Electric power-generating system using magnetic coupling for deeply implanted medical electronic devices. IEEE Trans. Magn. **38**(5), 3006–3008 (2002)
5. A.P. Sample, D.J. Yeager, P.S. Powledge, J.R. Smith, Design of a passively-powered, programmable sensing platform for UHF RFID systems, in *IEEE International Conference on RFID, 2007* (2007), pp. 149–156
6. M. Kishi, H. Nemoto, T. Hamao, M. Yamamoto, S. Sudou, M. Mandai, S. Yamamoto, Micro thermoelectric modules and their application to wristwatches as an energy source, in *Eighteenth International Conference on Thermoelectrics, 1999 (1999)*, pp. 301–307
7. R. Venkatasubramanian, C. Watkins, D. Stokes, J. Posthill, C. Caylor, Energy harvesting for electronics with thermoelectric devices using nanoscale materials, in *IEEE International Electron Devices Meeting, 2007. IEDM 2007* (2007), pp. 367–370
8. M. Strasser, R. Aigner, C. Lauterbach, T.F. Sturm, M. Franosh, G. Wachutka, Micromachined CMOS thermoelectric generators as on-chip power supply, in *12th International Conference on TRANSDUCERS, Solid-State Sensors, Actuators and Microsystems, 2003*, vol. 1 (2003), pp. 45–48
9. P.D. Mitcheson, E.M. Yeatman, G.K. Rao, A.S. Holmes, T.C. Green, Energy harvesting from human and machine motion for wireless electronic devices. Proc. IEEE **96**(9), 1457–1486 (2008)
10. P.P. Mercier, A.C. Lysaght, S. Bandyopadhyay, A.P. Chandrakasan, K.M. Stankovic, Energy extraction from the biologic battery in the inner ear. Nat. Biotechnol. **30**(12), 1240–1243 (2012)
11. E.Y. Chow, C.-L. Yang, Y. Ouyang, A.L. Chlebowski, P.P. Irazoqui, W.J. Chappell, Wireless powering and the study of RF propagation through ocular tissue for development of implantable sensors. IEEE Trans. Antennas Propaga. **59**(6), 2379–2387 (2011)
12. J.S. Ho, S. Kim, A.S.Y. Poon, Midfield wireless powering for implantable systems. Proc. IEEE **101**(6), 1369–1378 (2013)
13. G. Yilmaz, O. Atasoy, C. Dehollain, Wireless energy and data transfer for in-vivo epileptic focus localization. IEEE Sens. J. **13**(11), 4172–4179 (2013)
14. C. Sauer, M. Stanacevic, G. Cauwenberghs, N. Thakor, Power harvesting and telemetry in CMOS for implanted devices. IEEE Trans. Circuits Syst. I: Regular Papers **52**(12), 2605–2613 (2005)
15. M. Catrysse, B. Hermans, R. Puers, An inductive power system with integrated bi-directional data-transmission. Sens. Actuators A: Phys. **115**(23), 221– 29 (2004) (The 17th European Conference on Solid-State Transducers)
16. F. Mazzilli, P.E. Thoppay, V. Praplan, C. Dehollain, Ultrasound energy harvesting system for deep implanted-medical-devices (IMDS), in *2012 IEEE International Symposium on Circuits and Systems (ISCAS)* (2012), pp. 2865–2868
17. K. Mathieson, J. Loudin, G. Goetz, P. Huie, L. Wang, T.I. Kamins, L. Galambos, R. Smith, J.S. Harris, A. Sher, D. Palanker, Photovoltaic retinal prosthesis with high pixel density. Nat Photonics **6**(12), 872–872 (2012)
18. P. Vaillancourt, A Djemouai, J.-F. Harvey, M. Sawan, EM radiation behavior upon biological tissues in a radio-frequency power transfer link for a cortical visual implant, in *Proceedings of the 19th Annual International Conference of the IEEE Engineering in Medicine and Biology Society, 1997*, vol. 6 (1997), pp. 2499–2502
19. E.G. Kilinc, F. Maloberti, C. Dehollain, Short-range remote powering for long-term implanted sensor systems in freely moving small animals, in *SENSORS, 2013 IEEE* (2013), pp. 1–4

20. K. Finkenzeller, *RFID Handbook: Fundamentals and Applications in Contactless Smart Cards and Identification*, 2nd edn. (Wiley, New York, 2003)
21. A. Kurs, A. Karalis, R. Moffatt, J.D. Joannopoulos, P. Fisher, M. Solja, Wireless power transfer via strongly coupled magnetic resonances. Science, **317**(5834), 83–86 (2007)
22. M. Kiani, U.-M. Jow, M. Ghovanloo, Design and optimization of a 3-coil inductive link for efficient wireless power transmission. IEEE Trans. Biomed. Circuits Syst. **5**(6), 579–591 (2011)
23. K.M. Silay, C. Dehollain, M. Declercq, Improvement of power efficiency of inductive links for implantable devices, in *Research in Microelectronics and Electronics, 2008. PRIME 2008. Ph.D.* (2008), pp. 229–232
24. K.M. Silay, D. Dondi, L. Larcher, M. Declercq, L. Benini, Y. Leblebici, C. Dehollain, Load optimization of an inductive power link for remote powering of biomedical implants, in *IEEE International Symposium on Circuits and Systems, 2009. ISCAS 2009* (2009), pp. 533–536
25. R.R. Harrison, Designing efficient inductive power links for implantable devices, in *IEEE International Symposium on Circuits and Systems, 2007. ISCAS 2007* (2007), pp. 2080–2083
26. M. Ghovanloo, S. Atluri, A wide-band power-efficient inductive wireless link for implantable microelectronic devices using multiple carriers. IEEE Trans. Circuits Syst. I: Regular Papers **54**(10), 2211–2221 (2007)
27. J. Kim, H. Kim, K.D. Pedrotti, Power-efficient inductive link optimization for implantable systems, in *2011 IEEE Radio and Wireless Symposium (RWS)* (2011), pp. 418–421
28. U.-M. Jow, M. Ghovanloo, Design and optimization of printed spiral coils for efficient transcutaneous inductive power transmission. IEEE Trans. Biomed. Circuits Syst. **1**(3), 193–202 (2007)
29. A.K. RamRakhyani, S. Mirabbasi, M. Chiao, Design and optimization of resonance-based efficient wireless power delivery systems for biomedical implants. IEEE Trans. Biomed. Circuits Syst. **5**(1), 48–63 (2011)
30. A.P. Sample, D.A Meyer, J.R. Smith, Analysis, experimental results, and range adaptation of magnetically coupled resonators for wireless power transfer. IEEE Trans. Ind. Electron. **58**(2), 544–554 (2011)
31. G. Yilmaz, C. Dehollain, An efficient wireless power link for implanted biomedical devices via resonant inductive coupling, in *2012 IEEE Radio and Wireless Symposium (RWS)* (2012), pp. 235–238
32. K.M. Silay, Remotely Powered Wireless Cortical Implants for Brain-Machine Interfaces. Ph.D. thesis, EPFL
33. S.S. Mohan, M. del Mar Hershenson, S.P. Boyd, T.H. Lee, Simple accurate expressions for planar spiral inductances. IEEE J. Solid-State Circuits **34**(10), 1419–1424 (1999)
34. G. Yilmaz, O. Atasoy, C. Dehollain, *Wireless Data and Power Transmission Aiming Intracranial Epilepsy Monitoring*, vol. 8765. (2013), pp. 87650D–87650D–8
35. G. Yilmaz, C. Dehollain, Single frequency wireless power transfer and full-duplex communication system for intracranial epilepsy monitoring. Microelectron. J. **45**(12), 1595–1602 (2014)
36. G. Yilmaz, C. Dehollain, A wireless power link for neural recording systems, in 2012 8th Conference on Ph.D. Research in Microelectronics and Electronics (PRIME) (2012), pp. 1–4
37. K.M. Silay, C. Dehollain, M. Declercq, Inductive power link for a wireless cortical implant with two-body packaging. IEEE Sens. J. **11**(11), 2825–2833 (2011)
38. P. Favrat, P. Deval, M.J. Declercq, A high-efficiency CMOS voltage doubler. IEEE J. Solid-State Circuits **33**(3), 410–416 (1998)
39. C.-L. Chen, K.-H. Chen, S.I. Liu, Efficiency-enhanced cmos rectifier for wireless telemetry. Electron. Lett. **43**(18), 976–978 (2007)
40. H. Chung, A. Radecki, N. Miura, H. Ishikuro, T. Kuroda. A 0.025–0.45 w 60 with 5-bit dual-frequency feedforward control for non-contact memory cards. IEEE J. Solid-State Circuits **47**(10), 2496–2504 (2012)
41. Y.-H. Lam, W.-H. Ki, C.Y. Tsui, Integrated low-loss cmos active rectifier for wirelessly powered devices. IEEE Trans. Circuits Syst. II: Express Briefs **53**(12), 1378–1382 (2006)
42. S. Guo, H. Lee, An efficiency-enhanced cmos rectifier with unbalanced-biased comparators for transcutaneous-powered high-current implants. IEEE J. Solid-State Circuits **44**(6), 1796–1804 (2009)

43. H.-M. Lee, M. Ghovanloo, An integrated power-efficient active rectifier with offset-controlled high speed comparators for inductively powered applications. IEEE Trans. Circuits Syst. I: Regular Papers **58**(8), 1749–1760 (2011)
44. H.-K. Cha, W.-T. Park, M. Je, A cmos rectifier with a cross-coupled latched comparator for wireless power transfer in biomedical applications. IEEE Trans. Circuits Syst. II: Express Briefs **59**(7), 409–413 (2012)
45. S.S. Hashemi, M. Sawan, Y. Savaria, A high-efficiency low-voltage CMOS rectifier for harvesting energy in implantable devices. IEEE Trans. Biomed. Circuits Syst. **6**(4), 326–335 (2012)
46. Y. Lu, W.-H. Ki, A 13.56 MHZ CMOS active rectifier with switched-offset and compensated biasing for biomedical wireless power transfer systems. IEEE Trans. Biomed. Circuits Systems **8**(3), 334–344 (2014)
47. K.M. Silay, C. Dehollain, M. Declercq, Numerical analysis of temperature elevation in the head due to power dissipation in a cortical implant, in *30th Annual International Conference of the IEEE Engineering in Medicine and Biology Society, 2008. EMBS 2008* (2008), pp. 951–956
48. A.M. Sodagar, K. Najafi, Extremely-wide-range supply-independent CMOS voltage references for telemetry-powering applications. Analog Integr. Circuits Signal Process. **46**(3), 253–261 (2006)
49. F. Silveira, D. Flandre, P.G.A. Jespers, A gm/id based methodology for the design of cmos analog circuits and its application to the synthesis of a silicon-on-insulator micropower ota. IEEE J. Solid-State Circuits **31**(9), 1314–1319 (1996)
50. A. Ayed, H. Ghariani, M. Samet, Design and optimization of CMOS OTA with GMID methodology using EKV model for RF frequency synthesizer application, in *12th IEEE International Conference on Electronics, Circuits and Systems, 2005. ICECS 2005* (2005), pp. 1–5
51. C.C. Enz, F. Krummenacher, E.A. Vittoz, An analytical MOS transistor model valid in all regions of operation and dedicated to low-voltage and low-current applications. Analog Integr Circuits Signal Process **8**(1), 83–114 (1995)
52. J. Guo, K. Nang Leung, A 6- μw chip-area-efficient output-capacitorless LDO in 90-nm CMOS technology. IEEE J. Solid-State Circuits **45**(9), 1896–1905 (2010)
53. P. Hazucha, T. Karnik, B.A. Bloechel, C. Parsons, D. Finan, S. Borkar, Area-efficient linear regulator with ultra-fast load regulation. IEEE J. Solid-State Circuits **40**(4), 933–940 (2005)
54. R.J. Milliken, J. Silva-Martinez, E. Sanchez-Sinencio, Full on-chip CMOS low-dropout voltage regulator. IEEE Trans. Circuits Syst. I: Regular Papers **54**(9), 1879–1890 (2007)
55. E.N.Y. Ho, P.K.T. Mok, A capacitor-less CMOS active feedback low-dropout regulator with slew-rate enhancement for portable on-chip application. IEEE Trans. Circuits Syst. II: Express Briefs **57**(2), 80–84 (2010)
56. P.Y. Or, K.N. Leung, An output-capacitorless low-dropout regulator with direct voltage-spike detection. IEEE J. Solid-State Circuits **45**(2), 458–466 (2010)
57. Y.-S. Hwang, M.-S. Lin, B.-H. Hwang, J.-J. Chen. A 0.35 $\times$ 03bc;m CMOS sub-1v low-quiescent-current low-dropout regulator, in *IEEE Asian Solid-State Circuits Conference, 2008. A-SSCC '08* (2008), pp. 153–156
58. C.-M. Chen, C.-C. Hung, A fast self-reacting capacitor-less low-dropout regulator, in *2011 Proceedings of the ESSCIRC (ESSCIRC)* (2011), pp. 375–378
59. S.S. Chong, P.K. Chan, A 0.9-/spl mu/a quiescent current output-capacitorless LDO regulator with adaptive power transistors in 65-nm CMOS. IEEE Trans. Circuits Syst. I: Regular Papers **60**(4), 1072–1081 (2013)
60. G. Yilmaz, C. Dehollain, Wireless energy and data transfer for neural recording and stimulation applications, in *2013 9th Conference on Ph.D. Research in Microelectronics and Electronics (PRIME)* (2013), pp. 209–212

Chapter 4
Wireless Data Communication

Abstract Risk factors associated with the transcutaneous wires employed for data transmission can be reduced by means of wireless data communication between external base station and the implant unit. It is worth noting that bidirectional communication is indispensable for the majority of the neural monitoring applications. For the sake of definition used throughout this book, transferring data from the external base station to the implant is called downlink, while data transmission in the reverse direction is called uplink communication. Downlink communication is commonly utilized to reprogram the implant chip. Possible benefits of using a configurable implant chip can be enlisted as choosing the recording channels, modification of sampling parameters, and parameters associated with data compression or signal processing. Uplink communication, on the other hand, carries the processed information acquired by the MEAs. Moreover, it may contain additional information related to power feedback to ensure maximum power transfer efficiency all the time. Note that the number of recording channels directly affects the power demand of the implant; therefore, the transmitted power should be arranged accordingly to minimize power dissipation in the implant. Possible schemes to realize bidirectional communication is to use a half-duplex communication on a single channel with the help of a multiplexing method or to realize full-duplex communication. Considering the current requirements of the neural implant applications, half-duplex communication which allows communication in only one direction instantaneously is sufficient. However, this does not imply that power and data transfer should be performed at a single frequency. Depending on the data rate requirements, for instance in the case of multi-unit activity recording, a second frequency could be utilized to increase the data rate and allow full-duplex communication.

G. Yılmaz and C. Dehollain, *Wireless Power Transfer and Data Communication for Neural Implants*, Analog Circuits and Signal Processing,
DOI 10.1007/978-3-319-49337-4_4

4.1 Bidirectional Wireless Communication

Numerous wireless data communication solutions have been proposed in the literature which can be classified into two groups: data communication on the power line by changing parameters of the wireless power transfer link [1–6] and employing a dedicated transceiver on both parts [7–12]. Downlink communication can be directly performed by modulating the signal source in amplitude or frequency [13]. Uplink communication performed by perturbing the characteristics of the power line is called as load modulation or load shift keying (LSK) for power transfer links based on magnetic coupling [14] and as backscattering for electromagnetic radiation-based links [15, 16]. Using a dedicated transceiver isolates power and data transmission channels so allows these two links to be designed independently [17]. Compromise between these two solutions can be done with respect to the power budget and data rate requirement of the specific recording application. However, it should also be noted that additional components such as antennas (for dedicated transceiver) will occupy a certain volume that may violate implant size restrictions. For both cases, selection of operation frequencies has to be done by considering the bandwidth requirement imposed by the data rate of the application.

Operation principle of load modulation can be explained as follows: Maximum power efficiency state for inductive link is achieved at resonance; therefore, deviating the system from the maximum efficiency will correspond to another state. If the switching between these two states (binary modulation scheme) is controlled with a data stream containing acquired and processed neural information, uplink communication is to be realized on the power line at the expense of reduced efficiency. Considering that zeroes and ones will be equiprobable for a long enough transmission, we expect the overall efficiency to be equal to the average of the efficiencies of the two states.

Since resonant inductive links based on magnetic coupling are explained in detail in the remote powering section, here, load modulation schemes are discussed for the sake of completeness. As the name implies, the first versions of load modulation are constructed by changing the resistive load of the resonance tank [18]. Actually, changing any part of the resonance circuit actually degrades the efficiency and changes voltage and current waveforms in the external base station and, hence, corresponds to a change in the state. Next question is how to correlate the power transfer efficiency and the amount of deviation from maximum efficiency condition. Certain load modulation schemes, for instance short circuit or open circuit the resonance tank, therefore do not allow the transmitted power to reach to the load. In this case, we observe that efficiency is halved. However, is it really necessary to lose that amount of power to transmit data at that rate? This question aims to correlate data rate, bit-error rate, modulation index, and power transfer efficiency. More explicitly, higher modulation index on the implant side enables lower bit-error rate for a given data rate at the expense of reduced power transfer efficiency. In other words, an overdesign of the wireless communication blocks results in reduced performance in remote powering.

As the second solution, using a dedicated transmitter and a receiver is generally preferred when data rate is not achievable with load modulation. Note that the link bandwidth is bounded by the wireless power transfer frequency which is commonly selected as less than 10 MHz due to tissue absorbtion concerns [19]. Therefore, bandwidth could be widened, and hence, data rate could be increased by moving to a higher frequency. Briefly, in this solution, a carrier signal is generated and modulated with digitized neural information and by means of a transmitter antenna it is radiated. On the receiver side, an antenna picks up this signal and a front-end circuit (commonly composed of a low-noise amplifier, a mixer, and a local oscillator) amplifies and downconverts this signal to baseband.

4.2 Uplink Communication on the Power Transfer Link

Uplink communication is, as a part of single-frequency approach, performed on the power transfer link by detuning the resonance of the coils on the implant side. Hence, by using this technique, the resonance frequency of the load coil changes at the implant side, there will also be a reflected change at the external side which is used to recover modulated data.

Load modulation can be performed either at the input or output of the rectifier. The decision is a trade-off based on the power transfer efficiency and the data rate. More explicitly, higher data rate communication requires a wider bandwidth which results in a lower quality factor of the designed coil. Eventually, lower quality factor coils result in lower power transmission efficiency as explained in Sect. 3.3. Achieving a targeted data rate at a certain power transmission frequency becomes more difficult if load modulation is realized at the output of the rectifier than at the input of the rectifier. Note that an impedance variation at the output of the rectifier will first be translated into the input node and then reflected to the external side. Therefore, the modulation depth (or modulation index) will be lower than directly varying the impedance at the input node. Here it is important to note that as the modulation index gets lower, the data rate, which the demodulator at the external base station can support, decreases for a fixed bit-error rate. Nevertheless, it is worth noting that the impedance seen at the input of the rectifier is nonlinear and time-varying due to the switching of the pass transistor. In other words, the variation of the rectifier load impedance will not be linearly translated to the input which poses another limitation for the data rate.

Load modulation can be realized by changing (switching) any of the impedance parameters: R, L, or C. In the case of resonant inductive links, changing L or C has similar effects: Both change the resonance frequency of the implant coil. The power transfer efficiency obviously decreases when implant coil is detuned from the external coils' resonance frequency. Consequently, two binary states are defined if the resonance frequency of the implant coil is switched between this new frequency and maximum power transfer frequency. Since the capacitors possess a higher Q-factor than the inductors, it is more convenient to switch the capacitances in the practical applications in order to limit the power dissipation inside the implant. The third parameter which could be used for load modulation is R; however, changing R has certain drawbacks. First of all, it will decrease the Q-factor of the implant coil considering the Eq. 4.1:

$$Q = R\sqrt{\frac{C}{L}} = \frac{R}{\omega_0 L} \tag{4.1}$$

since the implant coil is a parallel resonant circuit. Note that added resistance will be in parallel with the parasitic resistance and definitely decrease equivalent resistance. Secondly, it will cause additional resistive loss inside the implant when additional R is conducting.

It may be claimed that changing L or C also decreases the power transfer efficiency, which is actually correct. At this point, it is important to remember the simple definition of efficiency: output power divided by input power. The mechanism reducing the efficiency in the case of L or C is different than the case of R: Switching L or C does not change the output power but necessitates an increase in the input power (transmitted power); however, addition of R to the implant coil increases the resistive loss inside the implant as well as reducing the Q-factor which necessitates an increase in the transmitted power. Therefore, changing C is superior than changing L or R for neural implants where power dissipation inside the implant is important.

Consequently, modulation of the power transmission frequency is performed by detuning the resonance frequency by adding a shunt capacitance to the parallel resonant circuit. The detuning capacitance is connected in series with the switching transistor. As a result, the resonance frequency on the implant side is switched according to the incoming bit stream. The method will henceforth be called as resonance frequency shift keying (RFSK).

Mechanism of modulation relies on the fact that power transfer efficiency is maximized when the coils are at resonance, and lower than this maximum point otherwise. More explicitly, a change in the resonance frequencies of the coils will disturb the system and alter the characteristics, namely the voltage on the coils. Therefore, connecting an additional capacitance (detuning capacitance) to the tuning capacitance of the load coil will detune the implant side of the inductive link and change the efficiency, as well as the voltage on the source coil. Therefore, it is possible to realize two levels, i.e., 0 and 1 of binary logic, by connecting the detuning capacitance in series with a switch which is controlled by the recorded neural information. Remark that the system has a certain bandwidth even without detuning, since the coils have finite Q-factors. Shifting the resonance frequencies increases the bandwidth at the expense of reduced power transfer efficiency.

4.2.1 Modulator

The load modulation circuit for uplink communication, depicted in Fig. 4.1, is designed and realized on silicon using UMC 180 nm MM/RF technology. The PMOS transistor which is employed as a switch and the series detuning capacitance constitutes the core of the modulator block. Additionally, the switching transistor is biased with dynamic bulk biasing method to prevent latch-up during operation. Note that highest potential oscillates between rectifier input and detuning capacitance nodes of the PMOS transistor. Furthermore, the gate of the PMOS is driven via an input buffer to enable switching at higher frequencies, more explicitly, at higher data rates.

Behavioral analysis of the modulator during operation can be summarized as follows: When the incoming data is 0, PMOS conducts and behaves as a resistor in series with the detuning capacitance. Consequently, the resonance frequency of the implant coil shifts and widens the bandwidth. Additionally, the quality factor of the parallel resonant circuit decreases due to ON resistance of the switching transistor. As an additional remark, traditional pass transistor scheme which is composed of a PMOS and a NMOS in parallel could not be used since dynamic biasing of NMOS requires triple-well process which is not available in some technologies. However, dynamic bulk biasing of PMOS has been already performed using gate-cross-connected PMOS transistors in standard p-substrate (n-well) process. In the

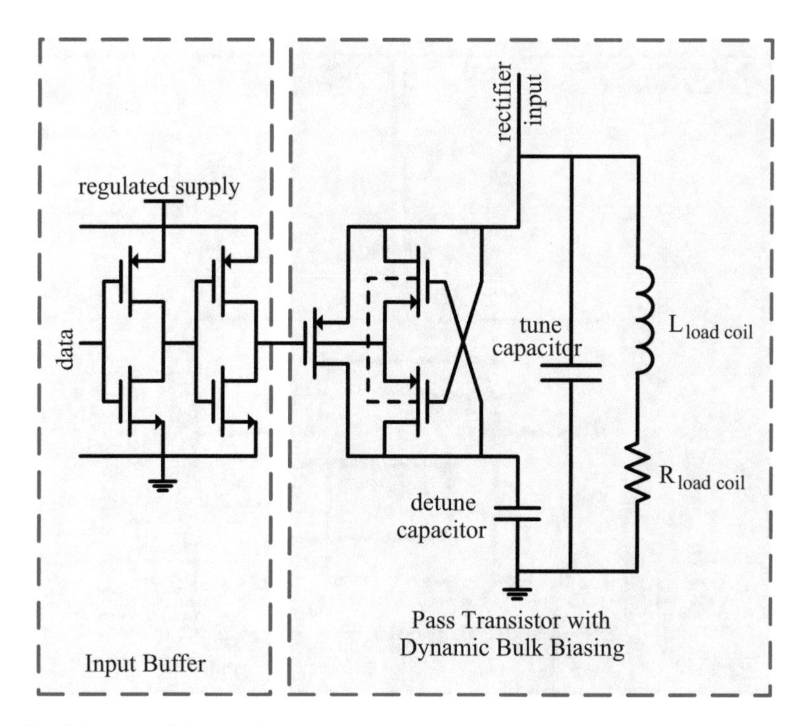

Fig. 4.1 Schematic of the modulator

other state, i.e., when data is 1 and PMOS is off, it behaves as a capacitance which is lower than 1 pF. Since the equivalent capacitance seen by the coil is almost equal to tuning capacitance (Eq. 4.2), the switch does not cause significant discrepancy in the maximum power transfer state.

$$C_{equivalent} = C_{tuning}||(C_{detuning} + C_{pmos}) \approx C_{tuning} \tag{4.2}$$

In order to characterize the switching performance of the modulator, we have measured the turn-on and turn-off times of the implemented modulator to be 16 and 52 ns, respectively. Therefore, maximum switching frequency of the modulator while maintaining 50% duty-cycle operation is found to be 9.6 MHz by using the Eq. 4.3 which is quite enough for uplink data rate aimed for fast ripple recording.

$$f_{maximum} = \frac{1}{max\{t_{ON}, t_{OFF}\}} \tag{4.3}$$

Since the detuning capacitance becomes effective when the switch is turned on, voltage drops when the switch is ON as depicted in Fig. 4.2. However, the achievable data rate is actually less than 9.6 Mbps (assuming maximum switching frequency is achieved at the modulator) since the modulation index seen by the demodulator input is smaller than the modulation index generated in the implant unit due to low

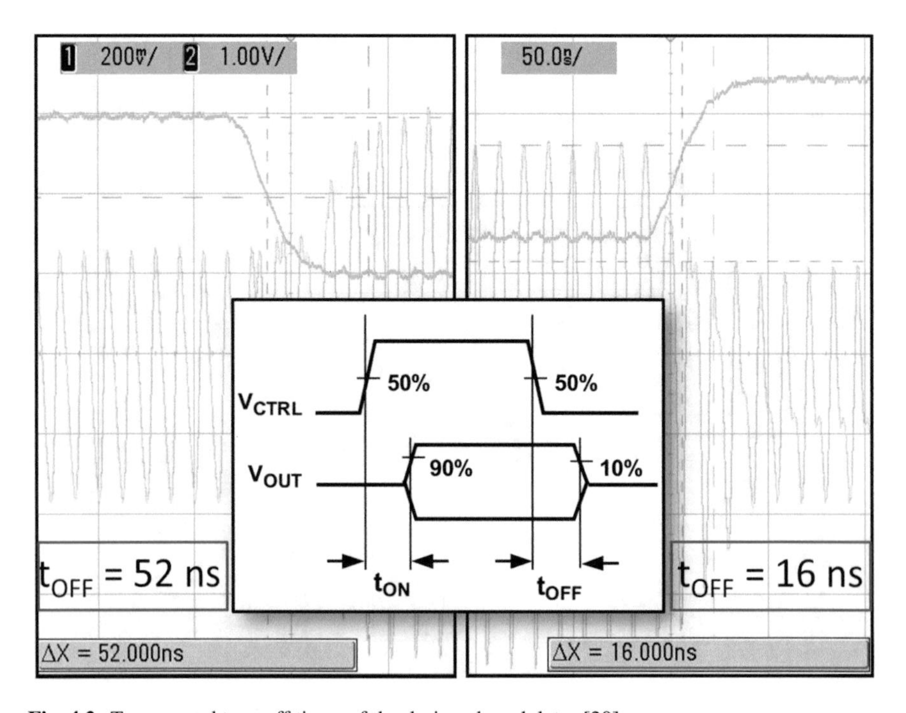

Fig. 4.2 Turn-on and turn-off times of the designed modulator [20]

coupling coefficient between the external base station and implant unit coils. For instance, any impedance on the implant side (Z_{imp}) is transformed to the external base station after being multiplied by k^2 as expressed in Eq. 4.4.

$$Z_{reflected} = k^2 Z_{imp} \tag{4.4}$$

This reflected impedance ($Z_{reflected}$) is added series to the impedance at the external base station (Z_{ext}); therefore, modulation index seen on the external base station (MI_{ext}) can be found by multiplying the one (MI_{imp}) on the implant side with the factor given in Eq. 4.5

$$MI_{ext} = MI_{imp} \frac{Z_{reflected}}{Z_{reflected} + Z_{ext}} \tag{4.5}$$

As the modulation index decreases, the demodulation becomes harder, so, in practice, data rate is further reduced to compensate the effect of reduced modulation index. Moreover, since it is not a data switch, but instead a detuning switch, its rail-to-rail operation is not critical. The critical operation parameter is its maximum switching frequency which sets the upper limit on the specifications for the data rate.

4.2.2 ASK Demodulator

Due to the coupling between implanted and external units, modulation in the implant is also observed on the external base station coils, but with a reduced modulation index. An envelope detector-based ASK demodulator is employed to demodulate this signal. Figure 4.3 presents the schematic of the demodulator which is designed by modifying the topology proposed in [21]. We have added a Schmitt-trigger instead of the output buffer employed in the original design. Therefore, we could eliminate glitches at the output of the amplifier.

Envelope detection is performed using a half-wave passive rectifier which is composed of a diode-connected PMOS transistor with dynamic bulk biasing and an on-chip metal-insulator-metal (MIM) capacitor. Since the modulation index is very low ($<4\%$), the difference between high and low states should be boosted or stretched. This amplification is performed by M_P which is self-biased via diode-connected M_D transistor. Operation principle of this amplifier can be explained via governing equations as follows.

When the input of the demodulator is high, and hence output of the envelope detector ($V_{envdetout}$), M_P will be turned on and working in saturation region, as well as M_D. Therefore, current passing through M_P and M_N can be written as in Eq. 4.6:

$$\begin{aligned} I_{M_P} = I_{M_N} &= \frac{\beta_P}{2}(V_{envdetout,high} - V_{Schtrin,high} + V_{\text{th}D} - |V_{\text{th}P}|)^2 \\ &= \frac{\beta_N}{2}(V_{Schtrin,high} - V_{\text{th}D} - V_{\text{th}N})^2 \end{aligned} \tag{4.6}$$

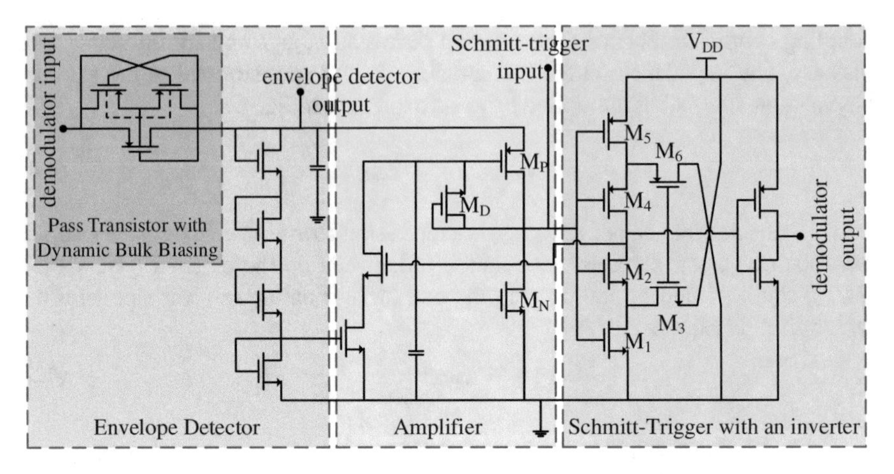

Fig. 4.3 Schematic of the envelop detector-based ASK demodulator [23]

where $V_{Schtrin,high}$ denotes the highest voltage level at the input of the Schmitt-trigger and V_{th} is the threshold voltage of the corresponding transistor.

Opposite situation where the input of the demodulator is low, and hence, the envelope detector output is low, M_P will be turned off, i.e., will be working in cutoff region. In this case, $V_{envdetout,low}$ can be expressed as in Eq. 4.7:

$$V_{envdetout,low} = V_{envdetout,high} + |V_{\text{th}P}| - V_{\text{th}D} \tag{4.7}$$

Returning back to the Eq. 4.6 and doing further simplification by assuming that M_N and M_D transistors have the same threshold voltage (Eq. 4.8), then $V_{envdetout,high}$ can be expressed as in Eq. 4.9

$$V_{\text{th}N} = V_{\text{th}D} = V_{\text{th,n}} \tag{4.8}$$

$$V_{envdetout,high} = V_{Schtrin,high} + |V_{\text{th}P}| - V_{\text{th,n}} + \sqrt{\frac{\beta_N}{\beta_P}}(V_{Schtrin,high} - 2V_{\text{th,n}}) \tag{4.9}$$

Following Schmitt-trigger (Fig. 4.3) eliminates the undesired fluctuations and drives the output buffer with a glitch-free signal. While designing the Schmitt-trigger, it is quite important to set low and high threshold values correctly [22]. Equations 4.10 and 4.11 give the design equations to set input high ($V_{\text{H}i}$) and input low ($V_{\text{L}i}$) voltages of the Schmitt-trigger where transitions occur [22]:

$$\frac{k_1}{k_3} = \left(\frac{V_{\text{DD}} - V_{\text{H}i}}{V_{\text{H}i} - V_{\text{TN}}}\right)^2 \tag{4.10}$$

$$\frac{k_4}{k_6} = \left(\frac{V_{\text{L}i}}{V_{\text{DD}} - V_{\text{L}i} - |V_{\text{TP}}|} \right)^2 \tag{4.11}$$

where

$$k = \mu C_{ox} \frac{W}{L} \tag{4.12}$$

for the corresponding transistor.

Otherwise bit durations on both sides do not match and the quality of communication decreases. This case can easily be observed if the supply voltage is set to a value other than 1.8 V. Therefore, the output of the Schmitt-trigger is set to be high when the value of the demodulator input is in between lowest and highest levels of the modulated signal. Therefore, the gate voltage of M_3 is slightly pulled down and consequently, M_3 conducts current as the demodulator input increases, and eventually changes the Schmitt-trigger output state to low. This variation turns off the pull-down transistor and allows the gate voltage of M_3 to increase. When the demodulator input starts to decrease, M_3 turns off and hence, Schmitt-trigger input decreases resulting a low-to-high transition at the output. In order to keep the logic convention between the demodulator input and output, an inverter is added to the output.

The demodulator dissipates less than 100 µW for 1 Mbps demodulation while driving an output load of 1 MΩ and 20 pF (impedance of the oscilloscope probe) in

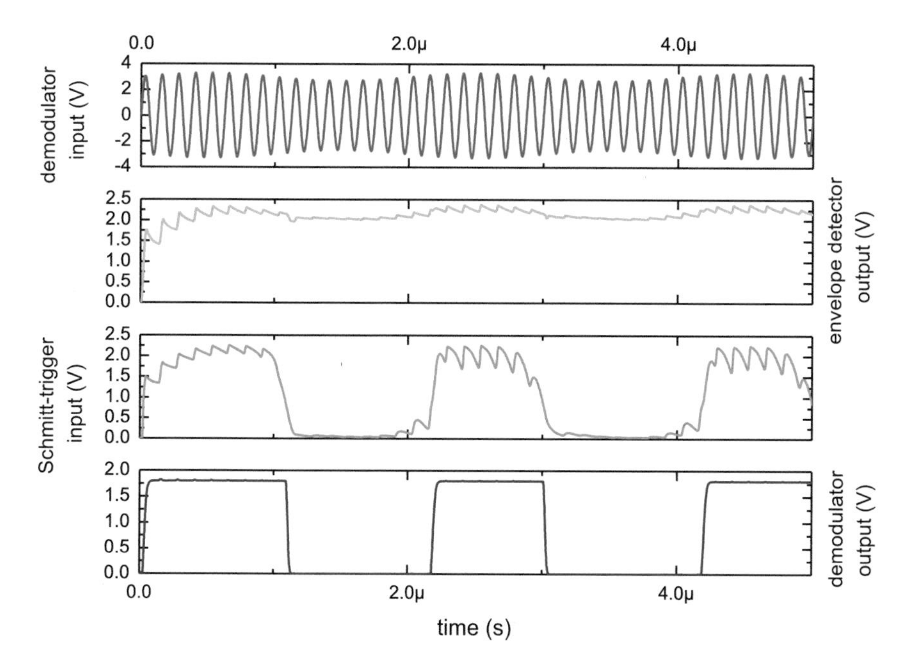

Fig. 4.4 Waveforms obtained from critical nodes of the demodulator through post-layout simulations while the operation frequency is 8 MHz, modulation frequency is 500 kHz, and modulation index is 10%

Table 4.1 Comparison of the designed ASK demodulator with similar works in the literature

Ref.	Process technology (nm)	Carrier frequency (MHz)	Data rate (kbps)	Modulation index (%)	Power consumption (μW)	Chip area μm^2	Highlights
[24]	350	13.56	1356	70	274	3555	ISM band
[25]	180	10	1370	NA	9000	1950000	High power and large area
[26]	350	2	10	NA	NA	20000	No capacitor
[27]	180	2	1000	50	336	468	No capacitor
[28]	350	2	250	27	1010	3025	No capacitor
[29]	3000	13.56	100	100	1600	225000	LTPS TFT process
[30]	250	20	256	100	15000	4100000	Complete receiver
[31]	180	2	18.7	NA	8.5	2411100	Complete system
[21]	350	13.56	1200	NA	306	3300	ISM band
This work	180	8.5	1500	4	100	12000	Self-sampled

the measurements. This power is supplied to the demodulator via 1.8 V DC supply. We can conclude that, compared to the power loss on the inductive link, power dissipation of the demodulator is negligible.

The benefit of using a Schmitt-trigger is more clear when Schmitt-trigger input node is observed in Fig. 4.4. The glitches at the output of the envelope detector are cleared thanks to the hysteresis characteristic of the Schmitt-trigger.

Table 4.1 gives a comparison of the designed ASK demodulator with similar structures presented in the literature. Main advantages of this work can be seen as working at very low modulation index values and having a self-sampled comparison mechanism. This allows the demodulator to work properly even when there is a variation in the absolute values of the incoming signal. On the other hand, it occupies a larger area compared to the capacitorless ASK demodulators. As a final remark, the designed demodulator could demodulate up to 1.5 Mbps in the experiments when the signal is supplied from a signal generator.

4.3 Uplink Communication with a Dedicated Transmitter and Receiver

Following the realization of the uplink communication for lower data rate link, another uplink communication is established using a second frequency in order to increase the data rate using an integrated transmitter in the implant and a discrete receiver on the external base station. Implant transmitter is composed of a voltage-controlled oscillator and a loop antenna. Receiver on the external base station consists of an antenna, low-noise amplifier, a mixer, a logarithmic amplifier, and a high-speed comparator [11]. Details of these circuits are explained in the following three subsections.

4.3.1 Oscillator

Carrier frequency is generated using a cross-coupled symmetrical voltage-controlled oscillator (VCO) whose schematic is presented in Fig. 4.5. This topology is chosen since it provides a balanced output and gives a better phase noise at a given power dissipation compared to the asymmetrical version [32]. The capacitance of the LC tank is implemented as a block of three capacitances for discrete tuning and a varactor for fine-tuning [11]. These capacitances are brought to resonance by using the inductance of a loop antenna instead of using an integrated inductance on chip. The antenna is directly connected to the differential output pins of the VCO. Therefore, it also serves as the transmitter antenna. Regarding the decision of loop antenna, it is worth noting that loop antennas exhibit a more robust performance closer to the body compared to its dual dipole antennas [33].

VCO is supplied through the output of the regulator (1.8 V). Measured average current during operation is 248 μA. While all of the discrete capacitors are shorted, the VCO can sweep a frequency band from 400 to 430 MHz thanks to the integrated varactor. Gain of the VCO (K_{VCO}) has been measured as 31.5 MHz/V in the linear region as shown in Fig. 4.6. This band covers MICS band (402–405 MHz) and two of the Medical Device Radiocommunications Service (MedRadio) bands (401 406 and 413 419 MHz) which enables utilization of the circuit in various medical communication channels.

In order to characterize the transmitter which is composed of a symmetrical cross-coupled LC VCO, a post-layout simulation has been run to find the transmitted power and efficiency. The electrical parameters of the loop antenna (for details: Sect. 4.3.2), namely inductance, loss resistance, and radiation resistance, have been inserted to the simulation environment. The oscillation frequency is fixed to the midpoint of the 413–419 MHz MedRadio band. Power consumed by the transmitter is calculated as 446 μW by taking the average of the supply current (248 μA) and multiplying it with

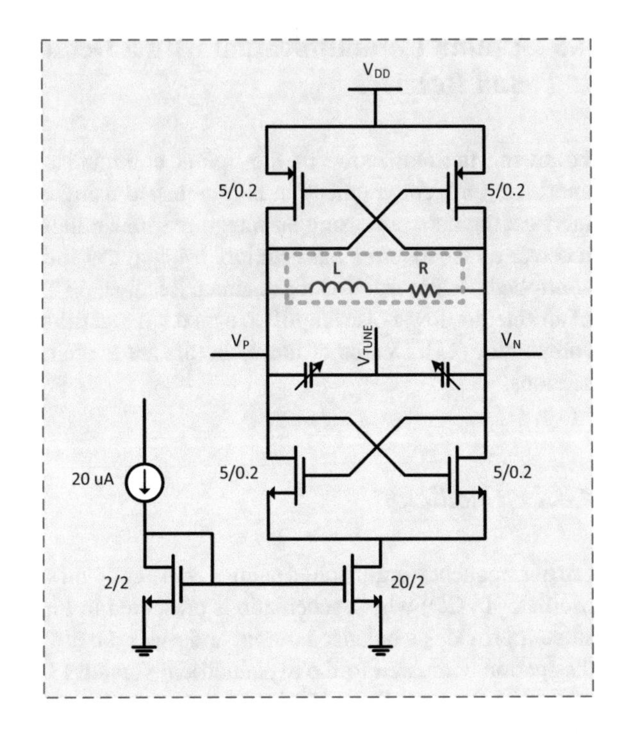

Fig. 4.5 Schematic of the LC cross-coupled voltage-controlled oscillator (dimensions given in μm). Note that inductance (L) and the series resistance (R) represent the off-chip antenna

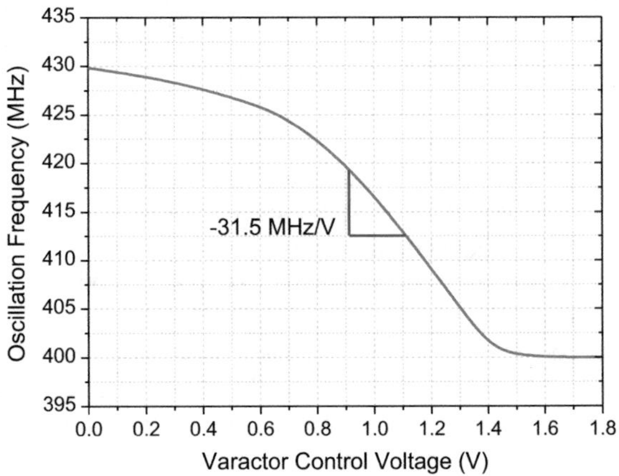

Fig. 4.6 Tuning range of the VCO frequency with respect to varactor control voltage

the value of the DC supply (1.8 V). Table 4.2 presents the transmitted power for the fundamental frequency (416 MHz) and second and third harmonics as well. According to these results, the efficiency of the transmitter is found as 4.2% (Eq. 4.13).

$$\eta_{transmitter} = \frac{P_{transmitted@fundamentalfrequency}}{P_{in,DC}} \tag{4.13}$$

Table 4.2 Transmitted power at the fundamental frequency, second harmonic, and third harmonic

Frequency	Transmitted power (dBm)	Transmitted power (μW)
Fundamental frequency (416 MHz)	−17.14	19.3
Second harmonic (832 MHz)	−98.48	$1.4\ 10^{-7}$
Third harmonic (1248 MHz)	−64.04	$3.9\ 10^{-4}$

Since the designed transmitter is not a conventional topology where we can measure the output power directly, I have measured the received power using the same antenna that will be used for external base station receiver (for details Sect. 4.3.3). The distance between the transmitter antenna and the receiver antenna is fixed to 60 cm and the 50 Ω receiver antenna (TI.10.0112) is connected to the spectrum analyzer. Figure 4.7 presents that the power level at 416 MHz is measured to be −61.7 dBm at the output of the receiver antenna. In addition, there is another strong signal visible which is exactly separated by the frequency of the wireless power transfer link. Since the external coils and the loop antenna are placed at a distance of 10 mm to each other, there is a direct coupling from the external coils to the loop antenna, which is also transmitted to the receiver. As the single-side bandwidth for the targeted data rate is almost one-fifth of the operation frequency of the wireless power transfer link, this undesired signal can be filtered out.

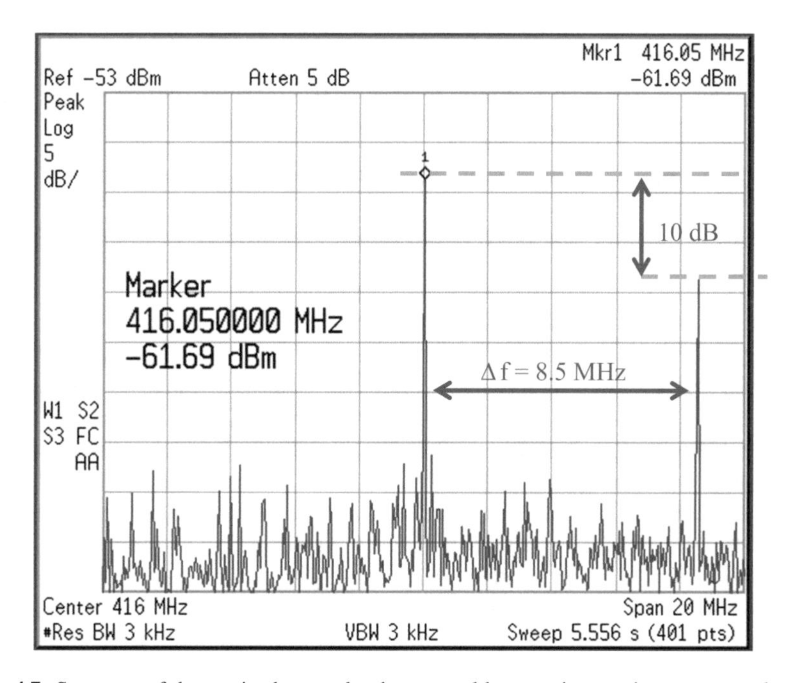

Fig. 4.7 Spectrum of the received power by the external base station receiver antenna when the radiation distance is fixed to 60 cm

4.3.2 Loop Antenna

In order to transmit the modulated carrier signal to the external base station, an antenna is employed which serves as a radiation source and an inductance for LC-coupled oscillator. Limited space for neural implants does not allow the utilization of resonant antennas at the targeted operation frequency (for instance, $f = 402\,\text{MHz}$, λ = 75 cm in air). In fact, the size of the antenna is limited to fit in a Burr hole with 15 mm diameter in this specific application. Such an antenna brings us to the definition of electrically small antennas.

Antennas with the largest dimension less than one-tenth of the wavelength are commonly referred as electrically small antennas (ESA) [34].

Electrically small loop antenna is defined as the dual of the short dipole antennas. Reactance of the loop antennas is inductive, whereas it is capacitive for dipole antennas. Since the capacitance of the oscillator is realized on the silicon substrate, exploiting the inductive reactance of the loop antenna is the major motive of using loop antenna.

Following the decision on the utilization of the loop antenna, geometrical design questions arise. Considering that these antennas are not the optimal solutions for a given frequency, a significant constraint is posed on the geometric design of the antenna in order to fit the antenna to the size specifications. At first glance, varying the shape of the loop antenna can be taken into account as a design parameter. However, for electrically small loop antennas, far field does not depend on the loop shape—whether it is square, circle, or elliptical [34]. Besides, considering that the radiation resistance can be increased by increasing the area of the loop, circular loop antenna stands out as the most proper solution since maximum area can be achieved if a circle is fit into another circle, namely Burr hole. Moreover, meandering the loop antenna does not improve the antenna efficiency since

1. current-carrying wires in opposite directions cancel out each other magnetic field and
2. increasing the total conductor length of the loop increases the loss resistance proportionally.

In cases where allowed implant space for the antenna allows a three-dimensional structure instead of a planar loop antenna on PCB, multi-turn loop antennas can be used to increase the radiation resistance. However, it is crucial to note that the resistance will be multiplied with the number of turns; in fact, it will increase more than that due to proximity effect combined with skin effect which already exists in both cases [35]. Consequently, radiation resistance can be improved by multi-turn loop antennas at the expense of reduced antenna efficiency. Furthermore, placing a permeable material, for instance ferrite, improves the radiation resistance and the antenna efficiency if the core material is chosen properly. Proper selection mainly requires a small loss tangent for the core material at the operation frequency so that additional loss due to the added instance does not cancel out its benefit on the radiation resistance. On the other hand, ferrite-loaded loops are not preferred for transmitter purposes since it causes nonlinearity and dissipation in the ferrite at high magnetic field strength is high.

Radiation resistance of an electrically small loop antenna is expressed as:

$$R_{radiation} = 320\ \pi^4\ \frac{A^2}{\lambda^4} \tag{4.14}$$

where A is the loop area and λ is the wavelength of the signal of interest [36].

Loss resistance of the loop antenna can be approximated as:

$$R_{loss} = \frac{P}{2w}\sqrt{\frac{\pi\ f\ \mu}{\sigma}} \tag{4.15}$$

where P corresponds to the perimeter of the loop, w is the track width on PCB, f is the operation frequency, μ is the magnetic permeability, and σ is the conductivity of the metal used on PCB [36].

Finally, inductance of the electrically small loop antenna is found from [36]:

$$L_{loop\ antenna} = \frac{\mu\ P}{2\pi\ \ln(\frac{8A}{wP})} \tag{4.16}$$

Design of the loop antenna is basically limited by the space allowed for the Burr hole implantation [37]. Therefore, we fix its mean diameter to 15 mm as shown in Fig. 4.8. Noting that increasing the width does not reduce the loss resistance of the antenna significantly due to skin effect, the width is fixed to 1 mm. Calculated loss resistance, radiation resistance, and inductance are 127 mΩ, 4.1 mΩ, and 32 nH, respectively [36]. Additionally, the designed loop antenna is simulated using CST Microwave Studio to analyze its electrical parameters and radiation characteristics such as gain and directivity of the antenna. Table 4.3 presents a comparison for the electrical parameters of the designed loop antenna between analytical calculations and numerical simulation results. Moreover, Fig. 4.10 presents three-dimensional (3D) directivity pattern of the designed loop antenna whose geometrical and physical definition in CST Microwave Studio is given in Fig. 4.9. Finally, Fig. 4.11 depicts the two-dimensional (2D) radiation pattern of the designed loop antenna. In summary, gain of the antenna is simulated to be −13.74 dB while maximum directivity is 1.76 dBi and radiation efficiency is −15.49 dB at 403 MHz, which falls into 402–405 MHz Medical Implant Communication Service (MICS) band.

Figure 4.8 shows the fabricated loop antenna on 0.3-mm-thick FR4 substrate and using 18-μm-thick copper (Cu) metallization with tin (Sn) finishing. Terminations of the loop antenna is mouse-bite cut and filled with metal to improve reliability of soldering for system integration.

Table 4.4 gives a comparison of the designed transmitter with similar works in the literature. It can bee seen that power consumption of the transmitter could be reduced by using an open-loop oscillator. Implemented transmitter exhibits a comparable performance with the recently published works using the similar frequency band.

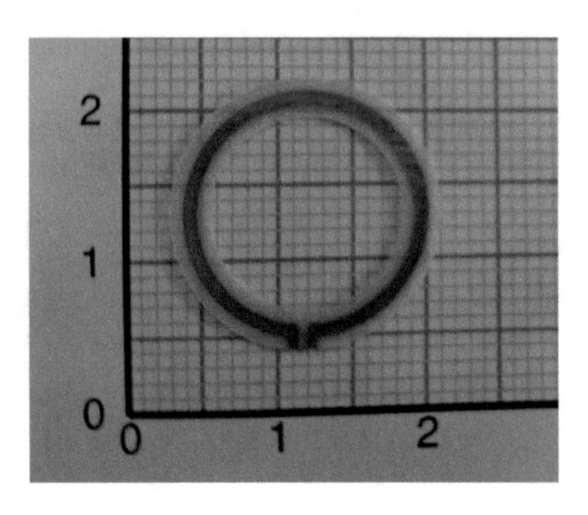

Fig. 4.8 Fabricated loop antenna which serves as both the inductance of the LC tank of the voltage-controlled oscillator and the antenna of the transmitter

Table 4.3 Analytically calculated and simulated (CST Microwave Studio) parameters of the designed loop antenna

Loop antenna parameters		
	Analytical	Simulation
Radiation resistance (R_rad)	4.1 mΩ	3.53 mΩ
Loss resistance (R_loss)	127 mΩ	125 mΩ
Radiation efficiency	−14.91 dB	−15.49 dB
Inductance	32 nH	35 nH

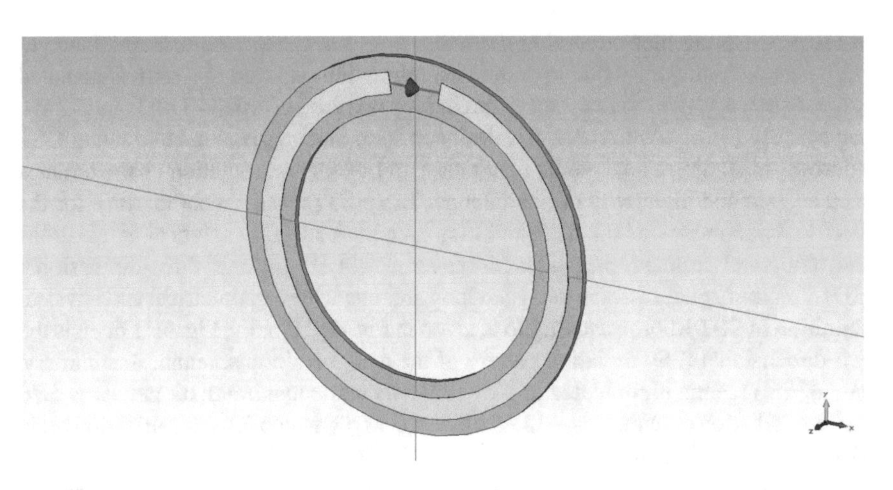

Fig. 4.9 Geometrical and physical definition of the designed loop antenna in CST Microwave Studio

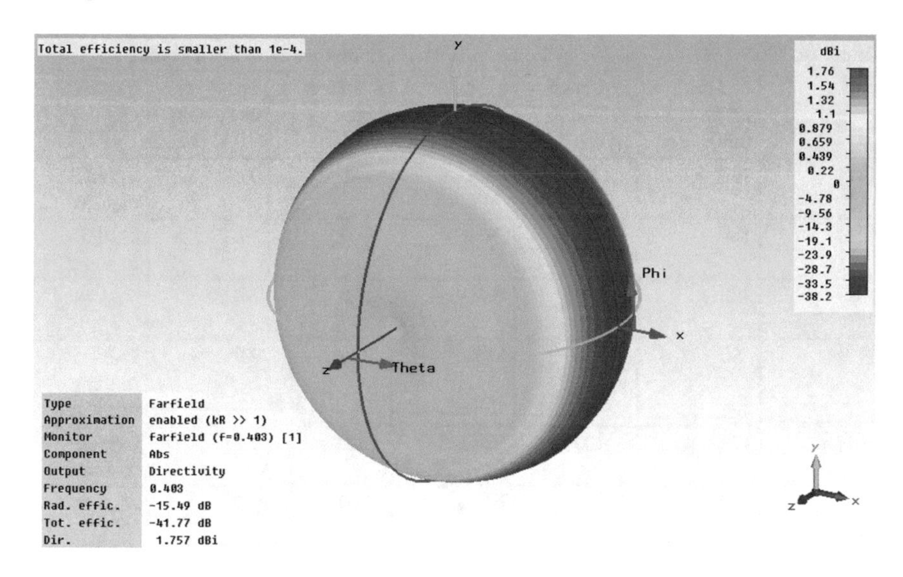

Fig. 4.10 3D directivity pattern of the designed loop antenna obtained from CST Microwave Studio

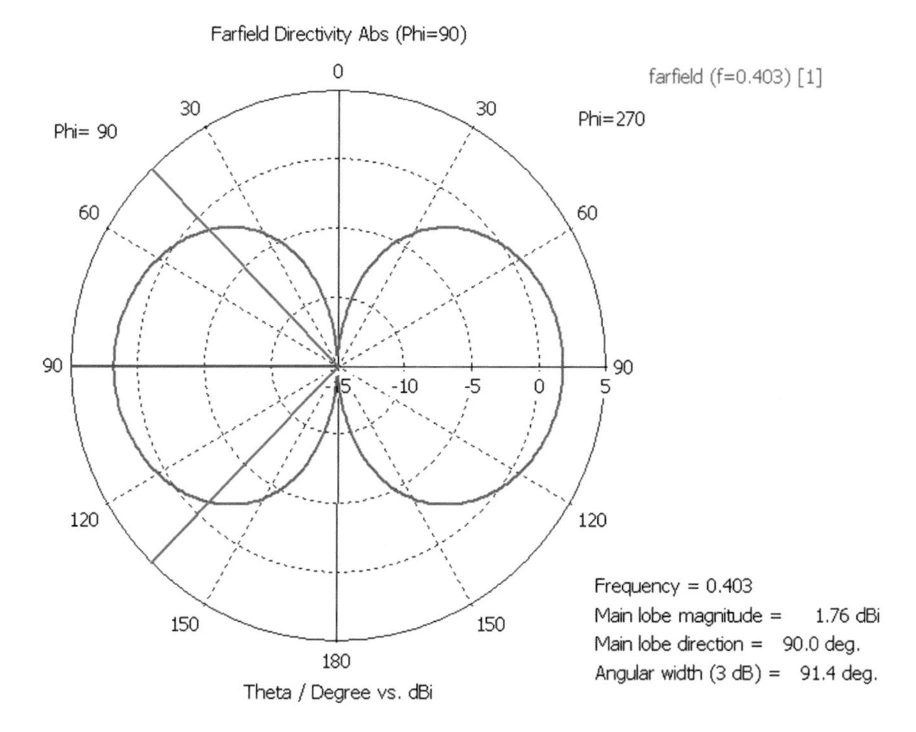

Fig. 4.11 2D radiation pattern of the designed loop antenna obtained from CST Microwave Studio

Table 4.4 Comparison of the designed transmitter with similar works in the literature

Ref.	Process technology (nm)	Power consumption (μW)	Modulation type	Data rate kbps	Energy per bit (nJ/bit)	Remarks
[17]	130	90	FSK	200	0.45	Crystal required
[38]	180	11550	OOK	800	14.43	TX/RX together
[39]	180	400	FSK	250	1.6	$V_{DD} = 0.7\,V$
[40]	90	350	MSK	120	2.9	Free-running oscillator
[41]	130	400	BFSK	100	4	Crystal required
[42]	180	8900	BFSK	1000	8.9	Inductorless VCO
This work	180	830	OOK	1800	0.46	Free-running oscillator

4.3.3 External Base Station Receiver

Receiver of the external base station for the uplink communication is built using off-the-shelf components since there is no size limitation. It is composed of a monopole antenna (TI.10.0112), a low-noise amplifier (ZFL-500, Mini-Circuits), a frequency mixer (ZX05-10, Mini-Circuits), a local oscillator (ZX95-535, Mini-Circuits), a logarithmic amplifier (AD8307), and a comparator (ADCMP601) as shown in Fig. 4.12 with corresponding specifications.

With this configuration of the receiver, the minimum detectable RF power at the input of the LNA is measured to be −85.5 dBm. This power level is found by fixing the mixer output power to −70 dBm and tracing back the signal power level on the receiver chain. −70 dBm is chosen according to the minimum detectable signal (MDS) limit of the logarithmic amplifier (AD8307) by leaving a 5 dB margin [43]. Figure 4.14 presents the power spectrum of the signal at the input of the LNA and Fig. 4.13 shows the power spectrum of the signal at the output of the mixer.

As an additional remark, leakage of the local oscillator (LO) to the mixer output can also be observed at the output of the logarithmic amplifier since LO signal is also in the bandwidth of the logarithmic amplifier. More explicitly, when −47 dBm RF power is present at the input of the LNA, power level of the downconverted IF signal becomes equal to the leakage of the LO signal, considering that the mixer provides a LO-RF isolation of 53 dB at 405 MHz [44] and LO has 6 dBm output power. The leakage of LO creates a DC offset at the output of the logarithmic amplifier. This offset problem can be solved either by using a filter to suppress the LO frequency at

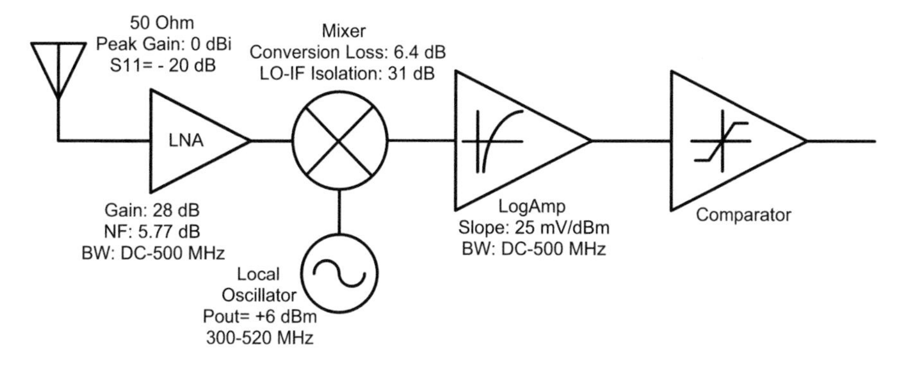

Fig. 4.12 Receiver topology and the specifications for the uplink communication channel

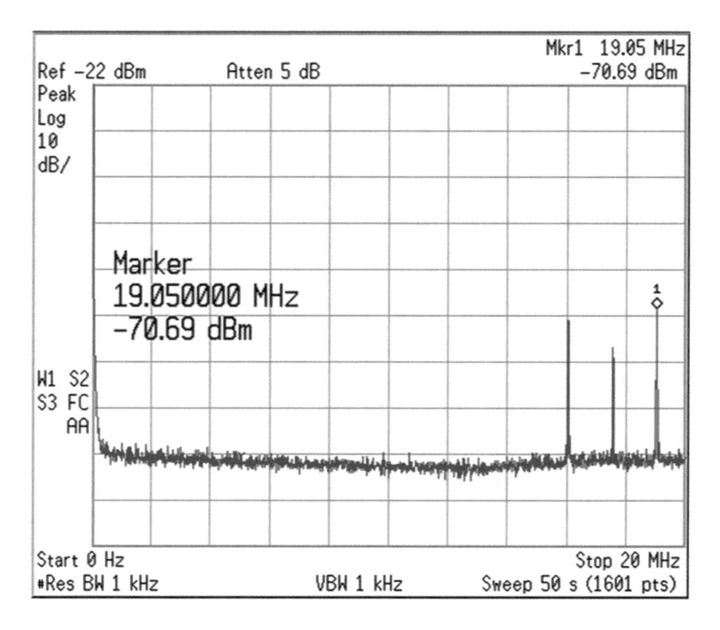

Fig. 4.13 Spectrum of the output of the mixer when its output power is fixed to -70 dBm

the output of the mixer or using a comparator at the output of the logarithmic amplifier. Here, I have implemented the latter solution using an off-the-shelf comparator (ADCMP601).

Since the output impedance of the mixer is 50 Ω, the input impedance of the logarithmic amplifier is brought to 50 Ω from 1.1 kΩ (the input resistance of AD8307) using a shunt 52.3 Ω. Therefore, impedance matching is achieved, which in return improves mixers driving capability. However, addition of a resistance, though it is shunt connected, brings additional noise source to the input of the logarithmic amplifier. Data sheet of AD8307 presents the input noise spectral density as 1.5 nV/$\sqrt{\text{Hz}}$ [43] and it will increase to 1.77 nV/$\sqrt{\text{Hz}}$ when the effect of thermal noise

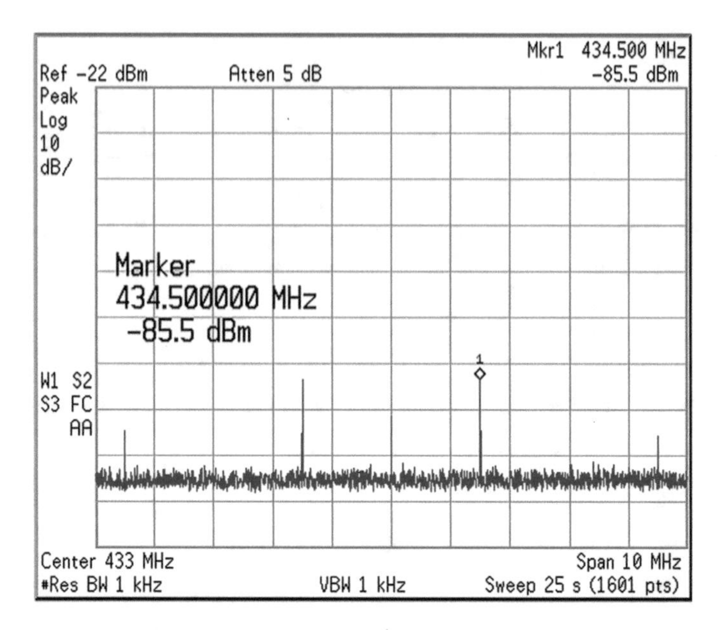

Fig. 4.14 Spectrum of the input of the low-noise amplifier when the output power of the mixer is fixed to −70 dBm

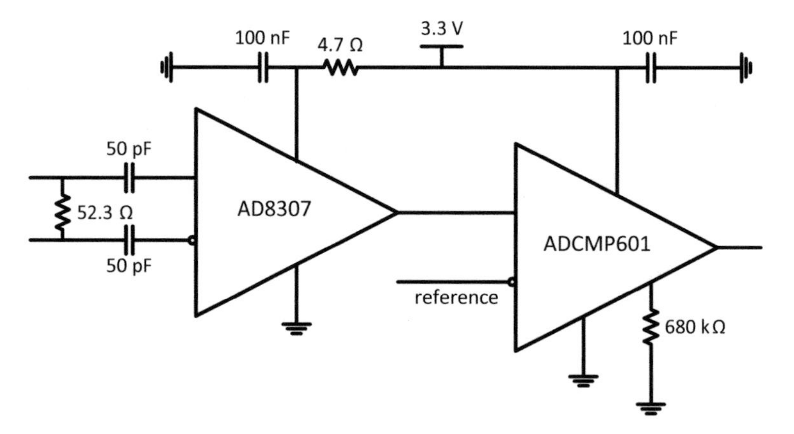

Fig. 4.15 Circuit schematic of the demodulator used for 433 MHz receiver. Minus polarity inputs of the logarithmic amplifier (AD8307) and the comparator (ADCMP601) are denoted with an inversion bubble

of the 52.3 Ω resistance is taken into account. However, this difference will be insignificant at the input of the low-noise amplifier (LNA) considering that it will be reduced by the gain of LNA, which is 28 dB. As the final remark, hysteresis pin of the high-speed comparator is terminated with 680 kΩ as shown in Fig. 4.15. Hysteresis voltage is lower than 10 mV with this control resistor value [45].

4.4 Downlink Communication

Downlink communication addresses the need for data transfer from the external base station to the implant in order to configure the sensor parameters as well as on-site processing parameters such as sampling rate or resolution of ADC. Downlink communication, in both system-level approaches, has been realized directly on the wireless power transmission link. Compared to the uplink communication performed by load modulation, it is fairly simpler due to the fact that coupling coefficient does not play a role for downlink communication. Remark that the variation in the impedance in the implant side is reflected to the external side with a scaling associated with the ratios of the inductances; however, a variation in the amplitude of the input signal will be directly seen on the implant unit without scaling of coupling coefficient. This eliminates the low modulation index problem encountered in uplink communication. Consequently, we could keep modulation index very small ($3\tilde{\%}$) in order not to deteriorate the wireless power transfer efficiency. Using a higher modulation index degrades the power transfer efficiency causing additional power dissipation in the implant, which in return increases the temperature more in the surrounding tissues.

The source signal, generated by a signal generator, is directly amplitude modulated using the embedded functions of the instruments. Demodulation on the implant side, however, is performed by using an ASK demodulator, which is, in fact, the same as the one used in uplink communication and explained in detail in Sect. 4.2.2 [13].

4.5 Clock Recovery

In addition to the communication circuits, a clock recovery circuit has been designed and implemented for the implant electronics. Note that acquired neural signals will be sampled and converted into digital signals. Therefore, either a clock generation in the chip or a clock recovery circuit is required. Here, we select the latter one for the sake of simplicity. Figure 4.16 presents the schematic of the clock recovery circuit which is composed of a Schmitt-trigger and a master-slave configuration D flip-flop (DFF). A square wave is generated at the half frequency of operation frequency by Schmitt-trigger and employing DFF ensures 50% duty cycle of the generated square wave.

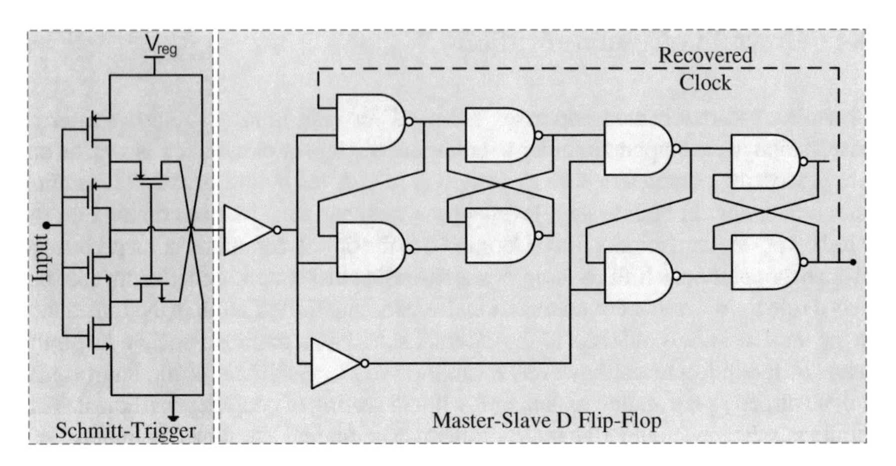

Fig. 4.16 Schematic of the clock recovery circuit

4.6 Summary

This chapter explains the wireless data communication part of the systems proposed in Chap. 2. First, two different approaches to realize bidirectional communication have been explained. Single-frequency approach aims to establish the bidirectional communication between the external base station and the implant on the same frequency as the wireless power transfer frequency. Corresponding modulation and demodulation circuits for uplink communication, as well as the load shift keying technique, have been explained. Regarding the two-frequency approach, a system-level information is followed by the design of the transmitter in the implant and the receiver in the external base station. The uplink transmitter is composed of an integrated free-running oscillator and a loop antenna, whereas the receiver is built using off-the-shelf components since there is no strict size limitation for that. Additionally, downlink communication and clock recovery circuits have been explained to conclude the chapter. Contents of this chapter have been published in [6, 11, 12, 20].

References

1. G. Wang, W. Liu, M. Sivaprakasam, G.A. Kendir, Design and analysis of an adaptive transcutaneous power telemetry for biomedical implants. IEEE Trans. Circ. Syst. I Regular Pap. **52**, 2109–2117 (2005)
2. M. Karimi, A.M. Sodagar, M.E. Mofrad, P. Amiri, Auxiliary-carrier load-shift keying for reverse data telemetry from biomedical implants, in *2012 IEEE Biomedical Circuits and Systems Conference (BioCAS)* (2012), pp. 220–223
3. Z. Tang, B. Smith, J.H. Schild, P.H. Peckham, Data transmission from an implantable biotelemeter by load-shift keying using circuit configuration modulator. IEEE Trans. Biomed. Eng. **42**(5), 524–528 (1995)

4. S. Sonkusale, Z. Luo, A complete data and power telemetry system utilizing bpsk and lsk signaling for biomedical implants, in *30th Annual International Conference of the IEEE Engineering in Medicine and Biology Society, 2008. EMBS 2008* (2008), pp. 3216–3219
5. G. Yilmaz, C. Dehollain, Capacitive detuning optimization for wireless uplink communication in neural implants, in *2013 5th IEEE International Workshop on Advances in Sensors and Interfaces (IWASI)* (2013), pp. 45–50
6. G. Yilmaz, O. Atasoy, C. Dehollain, Wireless energy and data transfer for in-vivo epileptic focus localization. Sensors J. IEEE **13**(11), 4172–4179 (2013). Nov
7. A. Tekin, M.R. Yuce, W, Liu. Integrated vco design for mics transceivers, in *Custom Integrated Circuits Conference, 2006. CICC'06. IEEE* (2006), pp. 765–768
8. F. Goodarzy, E. Skafidas, A fully integrated 200 x00b5;w, 40pj/b wireless transmitter for implanted medical devices and neural prostheses, in *2013 35th Annual International Conference of the IEEE Engineering in Medicine and Biology Society (EMBC)* (2013), pp. 3246–3249
9. J.L. Bohorquez, A.P. Chandrakasan, J.L. Dawson, A 350 μw cmos msk transmitter and 400 μw ook super-regenerative receiver for medical implant communications. IEEE J. Solid-State Circ. **44**(4), 1248–1259 (2009). April
10. M. Anis, M. Ortmanns, N. Wehn, Fully integrated uwb impulse transmitter and 402-to-405mhz super-regenerative receiver for medical implant devices, in *Proceedings of 2010 IEEE International Symposium on Circuits and Systems (ISCAS)* (2010), pp. 1213–1215
11. G. Yilmaz, C. Dehollain, Single frequency wireless power transfer and full-duplex communication system for intracranial epilepsy monitoring, in *2014 IEEE 12th International New Circuits and Systems Conference (NEWCAS)* (2014)
12. G. Yilmaz, C. Dehollain, An implantable system for intracranial neural recording applications, in *Biomedical Circuits and Systems Conference, 2014. BioCAS 2014. IEEE* (2014)
13. G. Yilmaz, C. Dehollain, Wireless energy and data transfer for neural recording and stimulation applications, in *2013 9th Conference on Ph.D. Research in Microelectronics and Electronics (PRIME)* (2013), pp. 209–212
14. S. Mandal, R. Sarpeshkar, Power-efficient impedance-modulation wireless data links for biomedical implants. IEEE Trans. Biomed. Circ. Syst. **2**(4), 301–315 (2008)
15. P.V. Nikitin, K.V.S. Rao, Theory and measurement of backscattering from rfid tags. IEEE Antennas Propag. Mag. **48**(6), 212–218 (2006)
16. K. Finkenzeller, *RFID Handbook: Fundamentals and Applications in Contactless Smart Cards and Identification*, 2nd edn. (Wiley, New York, 2003)
17. J. Pandey, B.P. Otis, A sub-100 μw mics/ism band transmitter based on injection-locking and frequency multiplication. IEEE J. Solid-State Circ. **46**(5), 1049–1058 (2011)
18. C. Sauer, M. Stanacevic, G. Cauwenberghs, N. Thakor, Power harvesting and telemetry in cmos for implanted devices, in *2004 IEEE International Workshop on Biomedical Circuits and Systems* (2004), pp. S1/8–S1–4
19. P. Vaillancourt, A Djemouai, J.-F. Harvey, and M. Sawan. Em radiation behavior upon biological tissues in a radio-frequency power transfer link for a cortical visual implant, in *Proceedings of the 19th Annual International Conference of the IEEE Engineering in Medicine and Biology Society, 1997*, vol. 6 (1997), pp. 2499–2502
20. G. Yilmaz, C. Dehollain, Single frequency wireless power transfer and full-duplex communication system for intracranial epilepsy monitoring. Microelectron. J. **45**, 1595–1602 (2014)
21. C.-C. Wang, C.-L. Chen, R.-C. Kuo, D. Shmilovitz, Self-sampled all-mos ask demodulator for lower ism band applications. IEEE Trans. Circ. Syst. II Express Briefs **57**(4), 265–269 (2010)
22. I.M. Filanovsky, H. Baltes, Cmos schmitt trigger design. IEEE Trans. Circ. Syst. I Fundam. Theory Appl. **41**(1), 46–49 (1994)
23. G. Yilmaz, O. Atasoy, C. Dehollain, Wireless data and power transmission aiming intracranial epilepsy monitoring, vol. 8765 (2013), pp. 87650D–87650D–8
24. J. Kim, K. Pedrotti, 202pj/bit area-efficient ask demodulator for high-density visual prostheses. Electron. Lett. **48**(9), 477–479 (2012)
25. B. Chi, J. Yao, P. Chiang, Z. Wang, A fast-settling wideband-if ask baseband circuit for a wireless endoscope capsule. IEEE Trans. Circ. Syst. II Express Briefs **56**(4), 275–279 (2009)

26. C.-C. Wang, T.-J. Lee, Y.-T. Hsiao, U. Fat Chio, C.-C. Huang, J.-J.J. Chin, Y.-H. Hsueh, A multiparameter implantable microstimulator soc. IEEE Trans. Very Large Scale Integr. (VLSI) Syst. **13**(12), 1399–1402 (2005)
27. C.-S.A. Gong, M.-T. Shiue, K.-W. Yao, T.-Y. Chen, Y. Chang, C.-H. Su, A truly low-cost high-efficiency ask demodulator based on self-sampling scheme for bioimplantable applications. IEEE Trans. Circ. Syst. I Regular Pap. **55**(6), 1464–1477 (2008)
28. T.-J. Lee, C.-L. Lee, Y.-J. Ciou, C.-C. Huang, C.-C. Wang, All-mos ask demodulator for low-frequency applications. IEEE Trans. Circ. Syst. II Express Briefs **55**(5), 474–478 (2008)
29. Y. Yueh-Hua, Y.-J. Lee, Y.-H. Li, C.-H. Kuo, C.-H. Li, Y.-J. Hsieh, C.-T. Liu, Y.-J.E. Chen, An ltps tft demodulator for rfid tags embeddable on panel displays. IEEE Trans. Microw. Theory Techn. **57**(5), 1356–1361 (2009)
30. B. Chi, J. Yao, S. Han, X. Xie, G. Li, Z. Wang, Low-power transceiver analog front-end circuits for bidirectional high data rate wireless telemetry in medical endoscopy applications. IEEE Trans. Biomed. Eng. **54**(7), 1291–1299 (2007)
31. C.-H. Hsu, S.-B. Tseng, Y.-J. Hsieh, C.-C. Wang, One-time-implantable spinal cord stimulation system prototype. IEEE Trans. Biomed. Circ. Syst. **5**(5), 490–498 (2011)
32. A. Hajimiri, T.H. Lee, Design issues in cmos differential lc oscillators. IEEE J. Solid-State Circ. **34**(5), 717–724 (1999)
33. K. Fujimoto, J.R. James, *Mobile Antenna Systems Handbook* (Artech House, Norwood, 2001)
34. Electrically Small. Basic principles of electrically small antennas by gary breed editorial director (2007)
35. G.S. Smith, Proximity effect in systems of parallel conductors. J. Appl. Phys. **43**(5), 2196–2203 (1972)
36. G.S. Smith, Radiation efficiency of electrically small multiturn loop antennas. IEEE Trans. Antennas Propag. **20**(5), 656–657 (1972)
37. J.-T. Schantz, T.-C. Lim, C. Ning, S.H. Teoh, K.C. Tan, S.C. Wang, D.W. Hutmacher, Cranioplasty after trephination using a novel biodegradable burr hole cover: technical case report. Neurosurgery, 58(1 Suppl):ONS–E176; discussion ONS–E176, February 2006
38. P.D. Bradley, An ultra low power, high performance medical implant communication system (mics) transceiver for implantable devices, in *Biomedical Circuits and Systems Conference, 2006. BioCAS 2006. IEEE* (2006), pp. 158–161
39. J. Bae, N. Cho, H.-J. Yoo, A 490uw fully mics compatible fsk transceiver for implantable devices, in *2009 Symposium on VLSI Circuits* (2009), pp. 36–37
40. J.L. Bohorquez, A.P. Chandrakasan, J.L. Dawson, A 350 μ w cmos msk transmitter and 400 μ w ook super-regenerative receiver for medical implant communications. IEEE J. Solid-State Circ. **44**(4), 1248–1259 (2009)
41. S. Rai, J. Holleman, J.N. Pandey, F. Zhang, B. Otis, A 500 μw neural tag with 2 μ vrms afe and frequency-multiplying mics/ism fsk transmitter, in *IEEE International Solid-State Circuits Conference - Digest of Technical Papers, 2009. ISSCC 2009* (2009), pp. 212–213, 213a
42. K.-C. Liao, P.-S. Huang, W.-H. Chiu, T.-H. Lin, A 400-mhz/900-mhz/2.4-ghz multi-band fsk transmitter in 0.18- μm cmos, in *IEEE Asian Solid-State Circuits Conference, 2009. A-SSCC 2009* (2009), pp. 353–356
43. Analog Devices, Inc. Low cost, dc to 500 mhz, 92 db logarithmic amplifier, http://www.analog.com/static/imported-files/data_sheets/AD8307.pdf
44. Mini-Circuits. Coaxial frequency mixer, http://217.34.103.131/pdfs/ZX05-10L.pdf
45. Analog Devices, Inc. Rail-to-rail, very fast, 2.5 v to 5.5 v, single-supply ttl/cmos comparators, http://www.analog.com/static/imported-files/data_sheets/ADCMP600_601_602.pdf

Chapter 5
Packaging of the Implant

Abstract Any kind of passive or active device which is to be implanted inside the body must satisfy certain biocompatibility and biosafety requirements and standards. It is worth noting that not harming the patient (*primum non nocere*) is one of the fundamentals of the medicine and medical sciences. These requirements have been established in years in order to prevent inflicting any damage to the human body. More explicitly, the implant may contain a toxic material which causes a direct damage to the tissues, or the body may refuse the implant even if the implant itself is not toxic, which causes an indirect damage. This refusal is called as foreign body reaction [1], and as a result, the immune system attacks the implant which results in an inflammation or swelling in return. In such cases, the implant should be removed immediately. Therefore, a set of preliminary experiments have to be conducted in order to get an approval. This chapter introduces the regulations and how these regulations are addressed within the frame of this work including a polymer-based packaging, and its modeling in terms of hermetical sealing is presented. Proposed packaging has been tested only for in vitro conditions and requires a detailed characterization to fulfill the requirements of biocompatibility before in vivo experiments.

5.1 Background

The part of the neural recording system which is to be implanted in the brain requires a packaging or encapsulation with certain features such as biocompatibility, biostability, and bidirectional diffusion barrier to maintain and sustain the operation of the implant properly. Note that the bare silicon chip will be recognized as a foreign body by the immune system which will immediately react to eliminate the threat. Since the immune system will create an inflammatory response, the implant will have to be removed with an additional surgery. Packaging the implant with a biocompatible material is, in fact, hiding it with a cap which is acceptable to the body.

G. Yılmaz and C. Dehollain, *Wireless Power Transfer and Data Communication for Neural Implants*, Analog Circuits and Signal Processing,
DOI 10.1007/978-3-319-49337-4_5

Device categorization by nature of body contact (see 5.2) – Category	Contact	Contact duration (see 5.3) A – limited (≤ 24 h) B- prolonged (>24 h to 30 d) C – permanent (> 30 d)	Cytotoxicity	Sensitization	Irritation or Intracutaneous reactivity	Systemic toxicity (acute)	Subchronic toxicity (subacute toxicity)	Genotoxicity	Implantation	Haemocompatibility
Surface device	Intact skin	A	X	X	X					
		B	X	X	X					
		C	X	X	X					
	Mucosal membrane	A	X	X	X					
		B	X	X	X	O	O		O	
		C	X	X	X	O	X	X	O	
	Breached or compromised surface	A	X	X	X	O				
		B	X	X	X	O	O		O	
		C	X	X	X	O	X	X	O	
External communicating device	Blood path, indirect	A	X	X	X	X				X
		B	X	X	X	X	O			X
		C	X	X	O	X	X	X	O	X
	Tissue/bone/dentin⁺	A	X	X	X	O				
		B	X	X	X	X	X	X	X	
		C	X	X	X	X	X	X	X	
	Circulating blood	A	X	X	X	X		O^		X
		B	X	X	X	X	X	X	X	X
		C	X	X	X	X	X	X	X	X
Implant device	Tissue/bone	A	X	X	X	O				
		B	X	X	X	X	X	X	X	
		C	X	X	X	X	X	X	X	
	Blood	A	X	X	X	X	X		X	X
		B	X	X	X	X	X	X	X	X
		C	X	X	X	X	X	X	X	X

Fig. 5.1 Initial evaluation tests for consideration in the frame of ISO 10993-1

An international standard (ISO 10993-1) is widely accepted as the guideline for biological evaluation and biocompatibility testing of medical devices. This standard defines the required tests to be conducted before implanting a biomedical device. Figure 5.1 shows the initial evaluation tests for consideration with respect to the device type, the type of the tissue which will be in contact, and the duration of the implantation.

According to the contact duration, devices are categorized as follows:

- Limited exposure (less than 24 h)
- Prolonged exposure (24 h 30 days)
- Permanent contact (more than 30 days)

as also indicated in Fig. 5.1. Consequently, an implantable device which will monitor brain activity to detect epileptonegic tissues approximately for 2 weeks will be considered as a prolonged implant. A prolonged duration neural implant has to succeed from several tests regarding biological effects such as cytotoxicity, sensitization, genotoxicity, and implantation. This set of experiments, obviously, requires an elaborate study for each implant. In order to skip these steps, in this book, materials which are already proven to be biocompatible are exploited. It is critical to note that these materials will encapsulate the system; however, the mentioned biocompatibility tests approve the material only when the implant itself is formed using that material.

Therefore, here, we need to define another concept to extend the biocompatibility of the implant encapsulation material to the entire implant: bidirectional diffusion barrier properties. This term basically defines the quality of physical barrier between the implant and the surrounding medium, which is generally formed by tissues and bodily fluids. In an ideal case, there would be no mass transport between these two media.

Note that the tissue interaction takes place on the interface, so the body will treat the implant as if it were completely composed of the packaging material if the encapsulation prevents any leakage from the implant. Here comes the importance of bidirectional diffusion barrier properties: Any material transport should be prevented in both directions, from the implant to the tissue medium and vice versa. One should keep in mind that the implant, for example, contains metals such as copper which is harmful to the body, and furthermore, surrounding body fluids are ionic and highly corrosive for the metals inside the implant. Another important feature of a successful packaging is biostability of the encapsulation material: It should also protect itself from the corrosive body fluids [2]. All of these three features have to be verified in vitro prior to in vivo operation. More explicitly, in vitro refers to studies and/or effects that occur in an artificial environment outside a living organism, whereas in vivo refers to studies and/or effects that occur within the body of living organisms. Testing the packaged implant inside the solutions that are mimicking body fluids such as mock cerebrospinal fluid (CSF) will give information about the diffusion barrier properties and biostability of the packaging material. This is, indeed, the method employed in this work for these tests, whose results are presented in the following section. It is clear that appropriate material selection for packaging is crucial to fulfill these three requirements as well as the practicalities of the target application.

Titanium and its alloys are commonly used for body implants such as cardiac pacemakers; however, it has two major drawbacks in intracranial applications:

1. It is not flexible and therefore not appropriate for large electrode arrays that have to conform the spherical shape of the brain.
2. It is not transparent to radio frequency (RF) which makes remote powering and wireless data communication impossible.

The latter behavior can be seen as a Faraday cage for the implant. Even in a scenario in which implant electronics are packaged separately from the inductive coils or antennae, RF signals induce eddy currents on the metallic surface of the casing and causes additional heat generation. In addition, separating implant electronics and inductive coils increases the area of the implant. At this point, polymers can replace titanium packages by overcoming these two issues [3]. Currently, polydimethylsiloxane (PDMS), poly(methyl methacrylate) (PMMA), polyimide, epoxy, and Parylene-C are the most studied polymers to package neural implants [4, 5]. Since they are transparent to both magnetic field and electromagnetic radiation, they enable lower frequency remote powering with higher frequency wireless data communication. Additionally, low-temperature cofired ceramic (LTCC) also possesses similar advantages and has an integration advantage in antenna fabrication. However, LTCC process is not widely available and lacks elasticity of the polymers. In addition

Table 3. Parylene Barrier Properties

Polymer	Gas Permeability at 25°C, (cc·mm)/(m²·day·atm)[a]				Water Vapor Transmission Rate (g·mm)/(m²·day)
	N_2	O_2	CO_2	H_2	
Parylene N	3.0	15.4	84.3	212.6	0.59[b]
Parylene C	0.4	2.8	3.0	43.3	0.08[c]
Parylene D	1.8	12.6	5.1	94.5	0.09[b]
Parylene HT	4.8	23.5	95.4	-	0.22[d]
Acrylic (AR)	-	-	-	-	13.9[e]
Epoxy (ER)	1.6	2.0 - 3.9	3.1	43.3	0.94[e]
Polyurethane (UR)	31.5	78.7	1,181	-	0.93 - 3.4[e]
Silicone (SR)	-	19,685	118,110	17,717	1.7 - 47.5[e]

a. ASTM D 1434
b. ASTM E 96 (90% RH, 37°C)
c. ASTM F 1249 (90% RH, 37°C)
d. ASTM F 1249 (100% RH, 38°C)
e. *Coating Materials for Electronic Applications*, Licari, J.J., Noyes Publications, New Jersey, 2003.

Fig. 5.2 Parylene barrier properties [19]

to the concerns on the whole body of the implant, one should consider same features for the electrode packaging and interconnection packaging for multiple sensor nodes. Obviously, the electrodes should be in touch with the tissue for recording. In terms of operational reliability, packaging should not disrupt the performance of recording electrodes. For electrode packaging, generally, noble metals such as platinum and gold are preferred to resist corrosion.

Upon deciding to use polymers for encapsulation, we have proposed a bilayer polymer package in this work to take advantage of different properties of polymers. Proposed package is composed of medical grade epoxy and Parylene-C, while it is Parylene-C which will be in contact with the tissues and body fluids. A cross-sectional view of the implant with bilayer polymer packaging is illustrated in Fig. 5.3. Parylene-C is a chemically resistant, hydrophobic material which is also pinhole free for depositions thicker than 1.5 nm [6]. Therefore, it has superb barrier properties, which is detailed in Sect. 5.2, and could be coated conformally at room temperature. However, it cannot be coated in the order of millimeters, but micrometers which decreases the time to fail. Nevertheless, Parylene-C has a finite vapor transmission rate, though smaller than others, as shown in Fig. 5.2, and the time during the implant can work safely is directly proportional to the thickness of the encapsulation. Therefore, firstly, a layer of biomedical grade epoxy is coated on top of the coils, integrated circuits, and other SMD components before coating Parylene-C.

Considering that the system will be implanted, it requires a biocompatible packaging to be accepted by the surrounding tissues; otherwise, it will result in an inflammation. On the other hand, hermetical sealing is enough for in vitro testing. As long as the packaging provides a bidirectional diffusion barrier between the solution and the implant, i.e., prevent chemical interaction, it can be used for preliminary experiments. It should be noted that sealing is important for both sides: Body fluids which are highly ionic may cause corrosion and other electrochemical reactions on the circuits and the inductors and possible leakage of non-biocompatible materials from the implant to the body tissues causes medical problems.

5.2 Diffusion Modeling of a Polymeric Package

The barrier properties of the system can be investigated via a modeling based on the mass transport from liquid phase to gas phase through a solid film. Since the permeability of the water vapor inside the bilayer polymer film is slower than the other transport mechanisms, transport from liquid to film and from film to gas, its resistance dominates the mass transport through the barrier film. Permeability is defined as [7] follows:

$$\text{Permeability} = \frac{\text{quantity of gas} \times \text{thickness}}{\text{area} \times \text{time} \times \text{partial pressure difference}} \tag{5.1}$$

In [7], permeability constant of water vapor through Parylene-C is calculated as $2 \times 10^{-9}\,\text{cm}^3$(STP)-cm/sec-cm^2-cmHg and the permeability constant for epoxy is given as 1.15×10^{-8} in [8]. Thickness of the film is the only design parameter that we are allowed to adjust in this application among the parameters affecting the permeability. Noting that bilayer polymer film is employed for encapsulation, equivalent permeability has to be calculated by taking the thickness of each film into consideration. Equivalent permeability of the bilayer polymer packaging can be approximated by equivalent conductance of series-connected conductances in an electrical circuit. Therefore, a threshold value for permeability/thickness can be defined to guarantee the operation of a given implant for a predefined duration.

In order to fabricate the package, we used a medical grade epoxy resin (thickness: $\approx$0.5 mm) to hermetically seal the coils and the electronics. On top of the epoxy, we have coated 2 μm Parylene-C which is already known to be biocompatible. Therefore, an equivalent permeability/thickness value of $2.25 \times 10^{-7}\,\text{cm}^3$(STP)/sec-cm^2-cmHg is achieved with this design. Note that the effective surface area of the encapsulation is 14 mm $\times$ 14 mm. Adding the thickness of the FR4 substrate (0.8 mm) to the coating thickness in both sides, we reach an implant thickness of 1.8 mm. Here, we have also employed vias and half cuts on the PCB to reinforce the adhesion of the epoxy and Parylene-C. Cross-sectional view of the packaging is illustrated in Fig. 5.3.

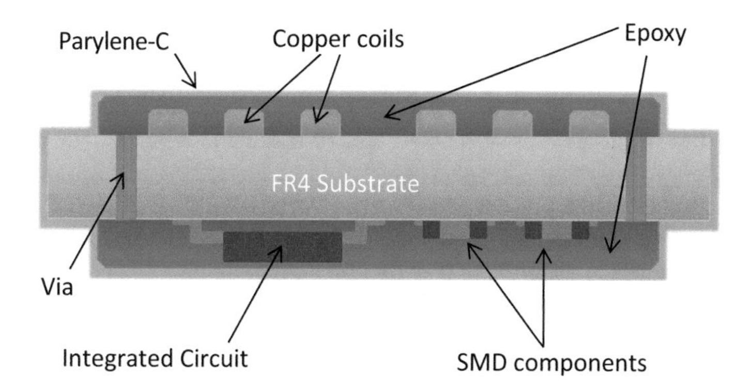

Fig. 5.3 Cross-sectional view of the packaged implant [9]

5.3 Temperature Elevation Considerations

Another critical operational issue for neural implants is the temperature elevation in the surrounding tissues due to heat generation as a result of implanting an active device. Although power dissipation in the implant contributes to the temperature elevation the most, the source of heat generation is not limited with it. Additional sources can be enlisted as remote powering and wireless data communication. Note that the waves generated for these two functions pass through the tissue and are absorbed at an amount depending on the frequency. This points out the significance of frequency selection for remote powering and wireless data communication. Moreover, power dissipation in the implant is not limited to the power dissipation of the circuits, but loss of the coils and eddy currents formed on nearby metals (e.g., electrodes) contributes to it. A specific discussion on heat generation mechanisms and recommendations for clearance of heat at the source is already given in [10] with an emphasis on brain–machine interfaces (BMI). In order to minimize the heat generation due to the electronic circuits, low-power circuit techniques are widely used in the literature for neural recording applications [11–14]. Moreover, high efficiency power transfer schemes are also studied to minimize the loss on the coils and the absorption on the tissues [15, 16]. Distribution of the components of the implant also improves the heat clearance as proposed in [10]. Silay et al. [17] show that two-body implants cause less temperature elevation than a single-body implant under same power dissipation. Although it is the minimum contributor to the heat generation, eddy current losses can also be eliminated with this method. Since its hard to define the boundary conditions and parameters in the analytical expression, temperature analysis are generally carried out by numerical methods [17]. Additionally, Kim et al. [18] presents a study on in vitro and in vivo experimental results of temperature elevation in the brain due to a neural implant by means of infrared imaging. As a final remark, using packaging does not change the steady-state conditions for temperature elevation in the surrounding tissue; however, it helps distributing the effect of localized heat sources.

5.4 Summary

In this section, we have given the details of the last major contribution of this work following the ones in the wireless power transmission and wireless data communication, namely as biocompatible packaging of a neural implant. Motivation behind a polymeric implant packaging has been explained, and the diffusion barrier properties of such a packaging material is modeled. Moreover, a brief explanation of temperature elevation issues in the brain in conjunction with packaging has been introduced. Content of this chapter has been published partially in [5].

Next chapter gives the results of the complete wireless power transfer and data communication system for different operation scenarios, including two different approaches in system-level and different working environments: in air, in mock CSF solution, and first stage on in vivo experiments on laboratory mice.

References

1. J.M. Anderson, A. Rodriguez, D.T. Chang, Foreign body reaction to biomaterials. Semin. Immunol. **20**(2), 86–100 (2008)
2. E. Dy, R. Vos, Jens Rip, A. La Manna, M.O. de Beeck, Biocompatibility assessment of advanced wafer-level based chip encapsulation, in *Electronic System-Integration Technology Conference (ESTC), 2010 3rd* (2010), pp. 1–4
3. C. Hassler, T. Boretius, T. Stieglitz, Polymers for neural implants. J. Polym. Sci. Part B Polym. Phys. **49**(1), 18–33 (2011)
4. K.M. Silay, C. Dehollain, M. Declercq, Inductive power link for a wireless cortical implant with two-body packaging. Sens. J. IEEE **11**(11), 2825–2833 (2011)
5. G. Yilmaz, O. Atasoy, C. Dehollain, Wireless energy and data transfer for in-vivo epileptic focus localization. Sens. J. IEEE **13**(11), 4172–4179 (2013)
6. J.-J. Senkevich, P.-I. Wang, Molecular layer chemistry via parylenes. Chem. Vapor Depos. **15**(4–6), 91–94 (2009)
7. M.A. Spivack, G. Ferrante, Determination of the water vapor permeability and continuity of ultrathin parylene membranes. J. Electrochem. Soc. **116**(11), 1592–1594 (1969)
8. D. Devanathan, R. Carr, Polymeric confonnal coatings for implantable electronic devices. IEEE Trans. Biomed. Eng. **27**(11), 671–674 (1980)
9. G. Yilmaz, O. Atasoy, C. Dehollain, *Wireless Data and Power Transmission Aiming Intracranial Epilepsy Monitoring*, vol. 8765 (2013), pp. 87650D–87650D-8
10. P.D. Wolf, Thermal considerations for the design of an implanted cortical brainmachine interface (BMI), in *Indwelling Neural Implants: Strategies for Contending with the in vivo Environment*, ed. by W.M. Reichert (CRC Press, Boca Raton, 2008)
11. W. Wattanapanitch, M. Fee, R. Sarpeshkar, An energy-efficient micropower neural recording amplifier. IEEE Trans. Biomed. Circuits Syst. **1**(2), 136–147 (2007)
12. J. Holleman, B. Otis, A sub-microwatt low-noise amplifier for neural recording, in *29th Annual International Conference of the IEEE Engineering in Medicine and Biology Society, 2007. EMBS 2007* (2007), pp. 3930–3933
13. R.R. Harrison, P.T. Watkins, R.J. Kier, R.O. Lovejoy, D.J. Black, B. Greger, F. Solzbacher, A low-power integrated circuit for a wireless 100-electrode neural recording system. IEEE J. Solid-State Circuits **42**(1), 123–133 (2007)
14. R. Sarpeshkar, W. Wattanapanitch, B.I Rapoport, S.K. Arfin, M.W. Baker, S. Mandal, M.S. Fee, S. Musallam, R.A Andersen, Low-power circuits for brain-machine interfaces, in *IEEE International Symposium on Circuits and Systems, 2007. ISCAS 2007* (2007), pp. 2068–2071
15. A.K. RamRakhyani, S. Mirabbasi, M. Chiao, Design and optimization of resonance-based efficient wireless power delivery systems for biomedical implants. IEEE Trans. Biomed. Circuits Syst. **5**(1), 48–63 (2011)
16. M. Kiani, U.-M. Jow, M. Ghovanloo, Design and optimization of a 3-coil inductive link for efficient wireless power transmission. IEEE Trans. Biomed. Circuits Syst. **5**(6), 579–591 (2011)
17. K.M. Silay, C. Dehollain, M. Declercq, Numerical analysis of temperature elevation in the head due to power dissipation in a cortical implant, in *Engineering in Medicine and Biology Society, 2008. EMBS 2008. 30th Annual International Conference of the IEEE* (2008), pp. 951–956
18. S. Kim, P. Tathireddy, R.A. Normann, F. Solzbacher, Thermal impact of an active 3-d microelectrode array implanted in the brain. IEEE Trans. Neural Syst. Rehabil. Eng. **15**(4), 493–501 (2007)
19. SCS Parylene Properties, http://scscoatings.com/wp-content/uploads/2015/09/parylene_properties.pdf

Chapter 6
System-Level Experiments and Results

Abstract This chapter presents the system-level integration of the proposed wireless power transfer and data communication system for the intracranial neural recording system. In the previous chapters, design and implementation of all the subblocks have been explained in detail. This chapter mainly focuses on the issues encountered during the integration and corresponding solutions while giving extensive experimental results. Experimental results include various combinations of functional blocks, namely wireless power transfer link, uplink communication, and data communication. Furthermore, it is worth nothing that the system has been experimentally tested in air, in vitro in a solution that mimics the cerebrospinal fluid (CSF), and in vivo on the cortex of a rat. More explicitly, the system has been completely characterized in air and in vitro; however, only the power management circuits have been employed for initial in vivo experiments.

6.1 System Integration and Characterization

The system blocks have been designed, fabricated, and characterized individually. Subsequently, the electronics is integrated with the inductive link and the antenna to enable wireless power transfer, uplink communication, and downlink communication. This section first describes the electrical characterization of the system using single frequency approach and then two-frequency approach. In vitro and in vivo experimental results are presented in the next section.

6.1.1 Single-Frequency Approach

Remarking the illustration given in Fig. 6.1, functional blocks, namely the wireless power transfer (WPT), uplink communication, and downlink communication circuits

G. Yılmaz and C. Dehollain, *Wireless Power Transfer and Data Communication for Neural Implants*, Analog Circuits and Signal Processing,
DOI 10.1007/978-3-319-49337-4_6

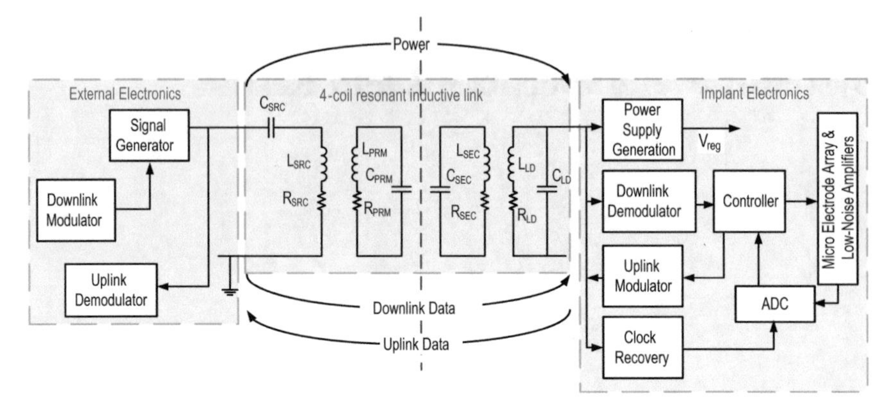

Fig. 6.1 System-level representation of the proposed single-frequency system composed of external electronics, 4-coil resonant inductive link, and implant electronics [1]

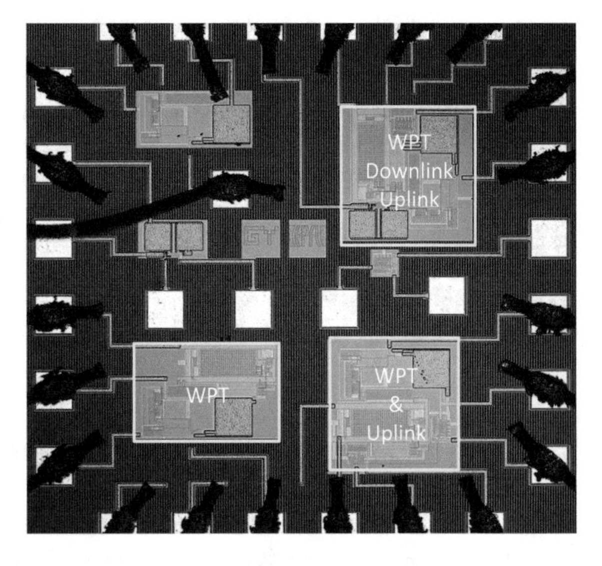

Fig. 6.2 Microscopic image of the integrated circuits used for characterization of only wireless power transfer (WPT); uplink communication with WPT; and bidirectional communication using single frequency with WPT

have been fabricated using UMC 180 nm 1P6M MM/RF process technology in different combinations of these functions in order to evaluate the performance metrics individually before the complete system evaluation. Figure 6.2 shows the microscopic image of the integrated circuits used for the characterization of only the wireless power transfer (WPT), uplink communication with WPT, and the bidirectional communication using single frequency with WPT. Additionally, Fig. 6.3 shows the micrograph of the integrated circuit wirebonded directly on the printed circuit board for electrical characterization purposes.

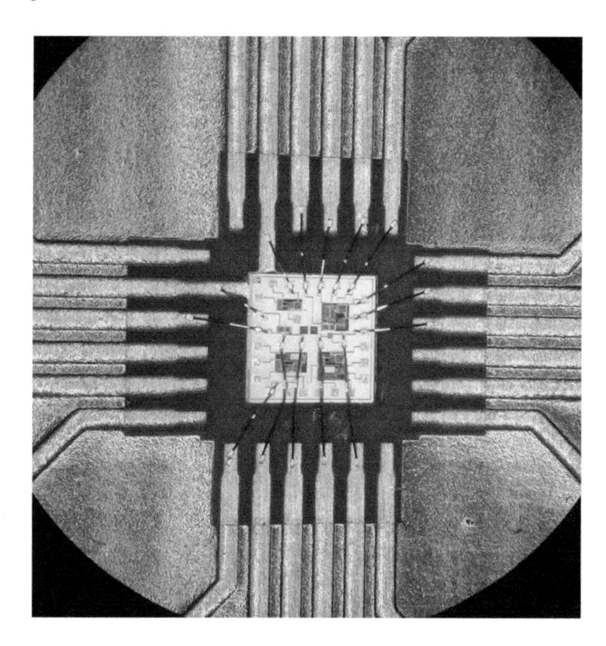

Fig. 6.3 Microscopic image of the integrated circuit (1st generation) wire-bonded directly on the printed circuit board (PCB)

6.1.1.1 Experiments

Figure 6.4 depicts the experimental setup employed to characterize the wireless power transfer and bidirectional data communication using the single-frequency approach.

The experimental characterization was first carried out for the wireless power transfer block as shown in Fig. 6.2. The source coil of the 4-coil resonant inductive link is driven by a signal generator (Agilent 33250A) at 8.5 MHz. Although geometric optimization was performed for 8 MHz, we have observed a variation due to the manufacturing and also found the maximum power transfer frequency by sweeping the proximate frequencies. The distances between the implant coils and the external coils are set to 10 mm, which is the average human skull thickness. The rectifier is connected to the load coil of the resonant inductive link, and combined efficiency is measured to be 46% while the rectifier output power is approximately 13 mW. After that, the regulator is connected and the overall power efficiency from the input of the inductive link to the regulator output has been measured as 36% for 10 mW output power and 1.8 V DC output voltage. Using the individual efficiencies of inductive link, the rectifier, and the regulator which are 55, 82, and 78%, respectively; overall power transfer efficiency can also be calculated as almost the same. In order to measure the input power, a series sense resistor (10 Ω) is used between the signal generator and the source coil of the inductive link. Power budget of 10 mW has been approximately calculated noting that there will be 4 recording sites which require 2.4 mW each as explained in Chap. 2.

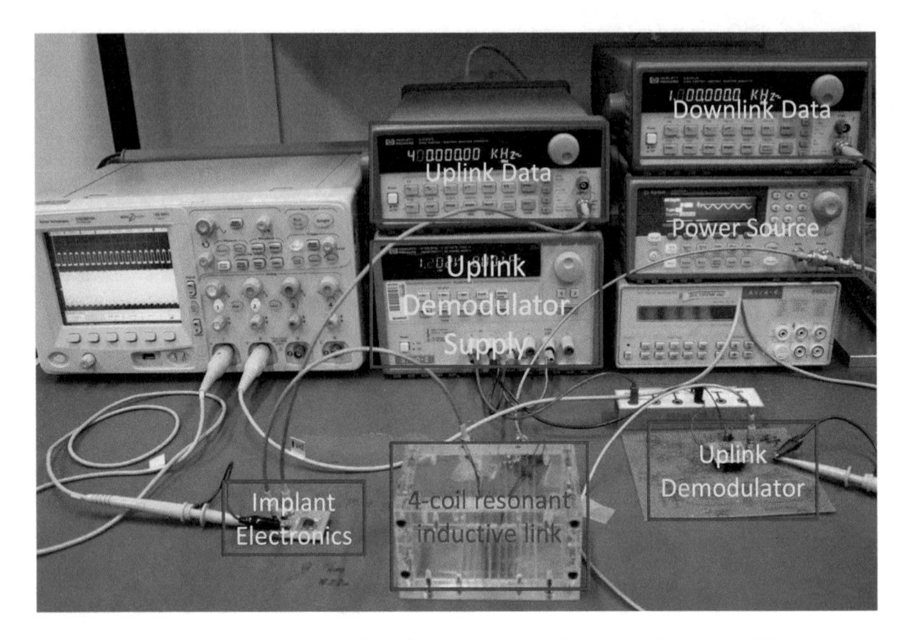

Fig. 6.4 Generic measurement setup for wireless energy and data transfer system [1]

Secondly, downlink communication is superposed on the wireless power transfer link. Since in a long bit stream 1s and 0s are equiprobable, a square wave at 50% duty cycle is generated and fed to the signal generator to modulate it in amplitude shift keying (ASK) mode as shown in Fig. 6.4. Although the downlink data rate demand for neural recording applications is generally less than 1 kbps, we have successfully tested the system upto 1 Mbps to see the limits. Figure 6.5 shows the downlink communication waveforms while signal generator is modulated with 1 MHz square wave. Consequently, power transfer efficiency while sustaining 1 Mbps downlink data transfer is measured to be 33%.

Thirdly, uplink communication is performed on the wireless power transfer link (without downlink communication). As explained in Sect. 4.2, a series transistor is switched on and off with an incoming square wave, emulating the digitized neural data, and hence resonance frequency of the load coil is detuned from the one of the external coils. Using a demodulator at the external base station, this variation is detected.

At this step, the most important parameter to be set is the detuning capacitance. Note that power transfer efficiency is maximum when both sides are at resonance; however, detuning means shifting the resonance frequency at the implant side, and hence, reducing the power transfer efficiency. On the other hand, the amount of shift defines the bandwidth which is the limiting factor of data rate. Therefore, the detuning capacitor should be chosen such that it creates enough bandwidth for 512 kbps data rate. If a larger capacitor is chosen, it costs an unnecessary reduction in the power transfer efficiency. However, it may also be required if the demodulator is not able

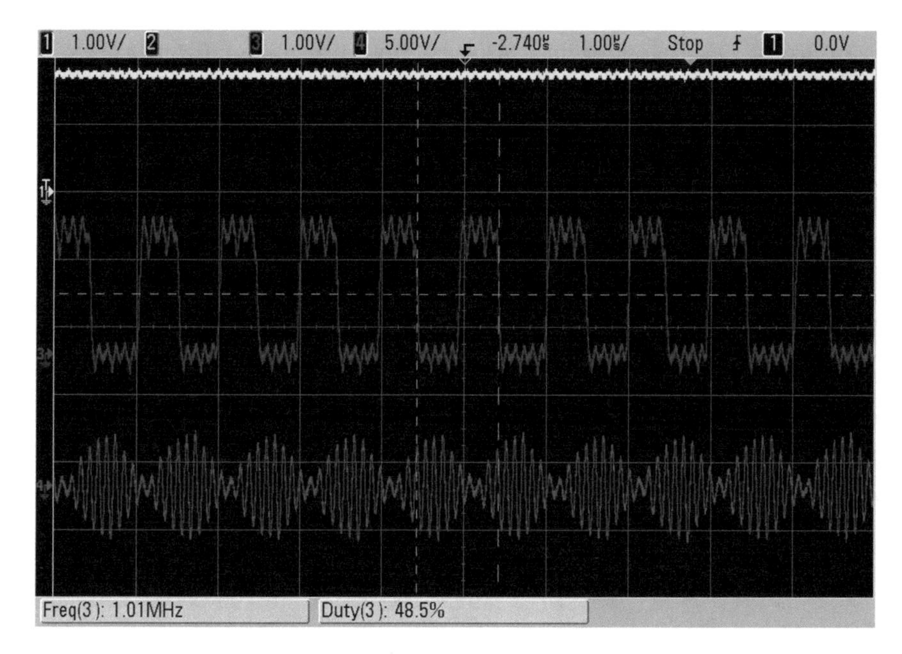

Fig. 6.5 Downlink communication waveforms at 1 MHz switching (*violet* demodulated downlink data and *pink* waveform at the terminals of the source coil)

to differentiate high and low signals at very low modulation index. In this case, 120 pF is found experimentally to be the appropriate detuning capacitance in terms of system-level functionality while maintaining the efficiency as high as possible.

Figure 6.6 depicts the power transfer efficiency with respect to data rates in air. In average, it can be concluded that power transfer efficiency drops to 33%. It is notable to see that power transfer efficiency barely changes with the data rate; however, as long as the duty cycle is preserved and the data rate is within limits, it is reasonable. Nevertheless the duty cycle defines the ratio of efficient and less-efficient power transmission, it is not related with the frequency of it upto the bandwidth limitation.

Following the realization of the wireless uplink communication, an experiment to optimize the detuning capacitance value with respect to energy per bit and power transfer efficiency has been conducted. As mentioned previously, increasing the detuning capacitance allows higher data rates since it enlarges the bandwidth, however, at the expense of the reduced power transfer efficiency. Moreover, increasing the detuning capacitance improves the modulation index for uplink: It increases from 2.1 to 4.4% when the detuning capacitance is increased from 120 to 330 pF. Decreasing power transfer efficiency with increasing detuning capacitance is plotted for 3 different detuning capacitances in Fig. 6.7 [1].

In order to keep the output power constant with the help of the regulator, the input power has to be increased accordingly. Therefore, we conclude that increased input power is responsible from the reduced efficiency, and this additional input power is dissipated for data transmission. Consequently, we divide this difference power to the bit rate to find the energy per bit. Figure 6.8 presents the energy per bit (E_b) values with respect to data rates for 3 different detuning capacitances [2]. From the point of optimization, for lower data rates, it is feasible to use a higher detuning capacitance if a better BER is required and to use a lower detuning capacitance if power transfer efficiency is critical. However, for data rates between 0.8 and 1 Mbps, the choice of capacitance becomes less important in terms of energy per bit which actually represents the limit of data communication for such configuration. Finally, careful observation of slopes indicates the existence of two separate regimes. Considering that slope is higher for lower data rates and lower for higher data rates, we conclude that a logarithmic function could be employed to model the relation between energy per bit and the data rate.

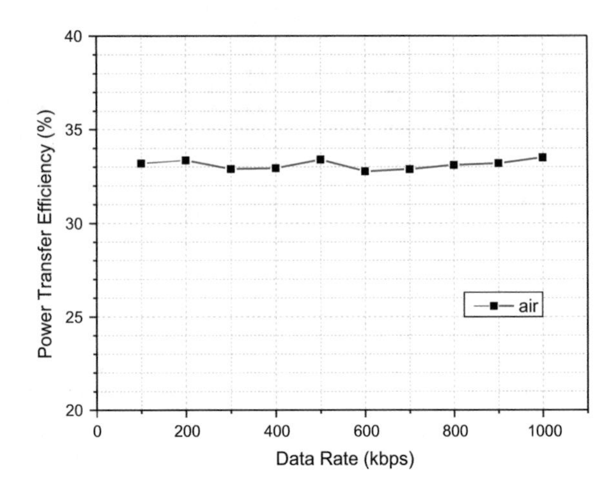

Fig. 6.6 Power transfer efficiency of the system with respect to uplink data rate in air

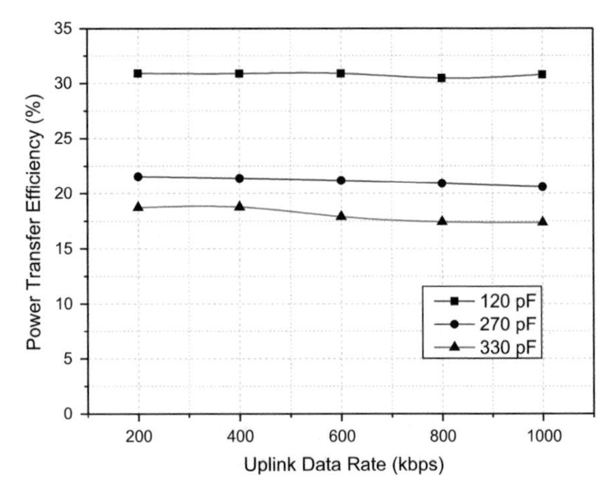

Fig. 6.7 Power transfer efficiency with respect to the uplink data rate for 3 different detuning capacitances [1]

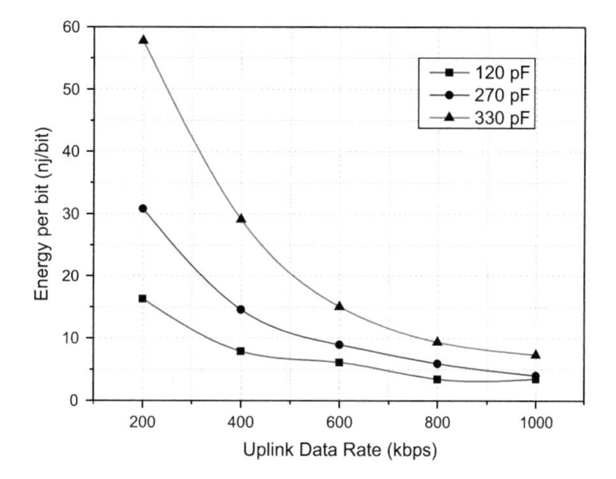

Fig. 6.8 Energy consumption per bit with respect to the uplink data rate for 3 different detuning capacitances [1]

Moreover, changing the modulation index affects the required signal-to-noise ratio (SNR) at the input of the demodulator, particularly at low MI levels. For instance, required SNR level changes from 13.75 to 10.5 dB when MI changes from 2.1 to 4.4% to have a detectable signal. It is assumed that the amplifier following the envelope detector gives very high gain to calculate these SNR levels. However, an increase in the data rate leads to an increase in the integrated noise bandwidth, meaning that the tolerable noise level in terms of V_{rms} decreases accordingly.

Finally, we have integrated these two unidirectional communication channels on the same frequency (or channel) as well as the wireless power link. As a result of this combination, we achieve a bidirectional data communication between the external and the implantable units. A micrograph of the implant electronics that perform these three functions, occupying a silicon area of $300 \times 350\ \mu m^2$, is shown in Fig. 6.9.

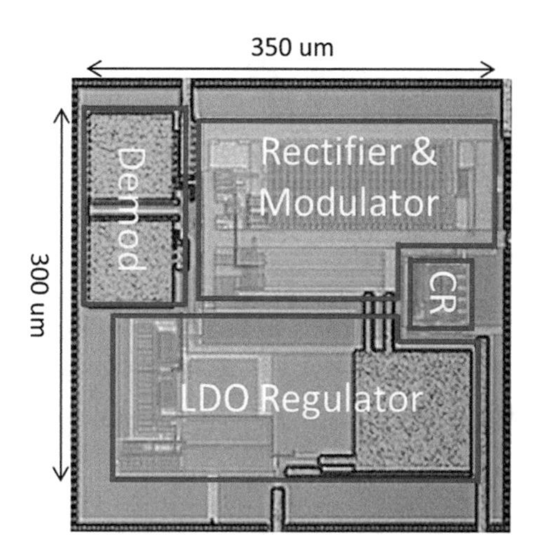

Fig. 6.9 Micrograph of the implant electronics composed of a rectifier, modulator, a low drop-out regulator, a demodulator (Demod), and a clock recovery circuit (CR) [1]

However, this integration has degraded the individual performances which are shown previously. As a result, fastest uplink communication can be performed while detuning is realized with a 400 kHz square wave at a 50% duty cycle. However, the shape of the output waveform (low and high durations) does not correspond to the one of the signal generators in terms of high and low durations. Therefore, it is necessary to employ a coding scheme in order to achieve a more reliable uplink communication. For instance, if well-known Manchester encoding, which actually halves the data rate, is employed, an uplink communication at 400 kbps data rate can be realized. It can be concluded that using a more efficient coding scheme will enable an uplink communication at 512 kbps. Figure 6.10 represents the waveforms captured from the input and output of the uplink demodulator as well as from the rectifier input (or from the terminals of the load coil) where the modulation is performed. In addition, downlink data rate is also limited by the uplink signal quality. We have reached a 1 kbps downlink communication while establishing a reliable 400 kbps uplink communication at the same time. Figure 6.11 represents the waveforms captured from the input and output of the downlink demodulator as well as the signal generator output where the modulation is performed. Downlink data rate is basically limited by the degradation in the uplink signal quality while performing full-duplex communication. The degradation becomes visible as the modulation index (MI) for downlink is increased while data rates are kept constant or the downlink data rate is

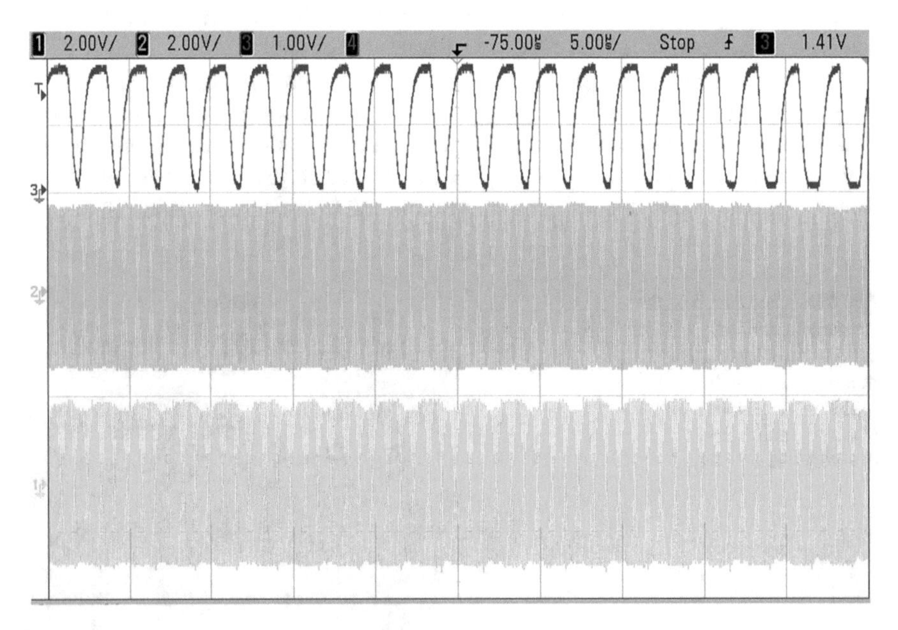

Fig. 6.10 Waveforms for 400 kpbs uplink communication; *violet* uplink demodulator output (1 V/div), *green* uplink demodulator input (2 V/div), and *yellow* rectifier input or load coil (2 V/div) (timescale: 10 μs/div)

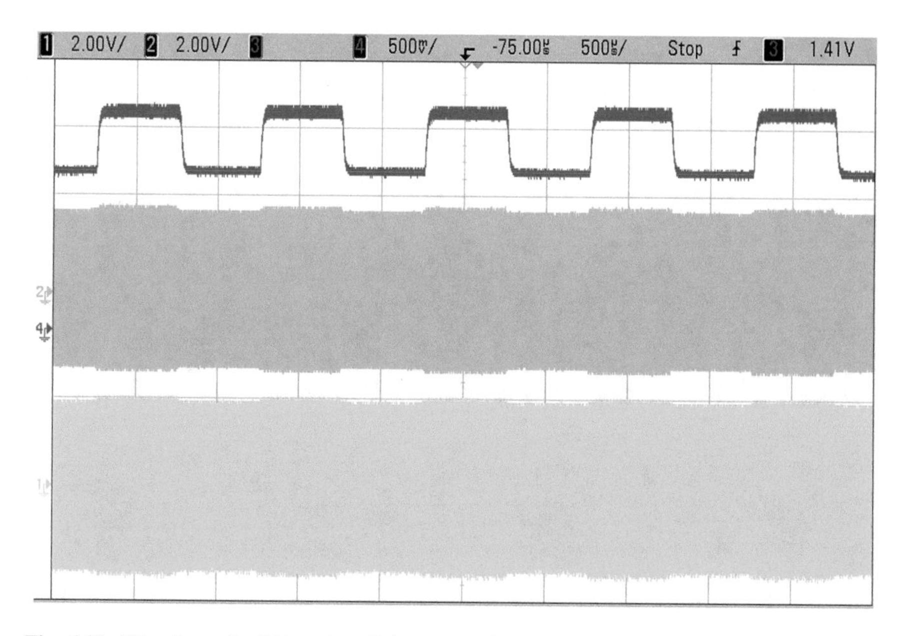

Fig. 6.11 Waveforms for 1 kbps downlink communication; *violet* downlink demodulator output (2 V/div), *green* signal generator output (2 V/div), and *yellow* downlink demodulator input (2 V/div) (timescale: 500 µs/div) [1]

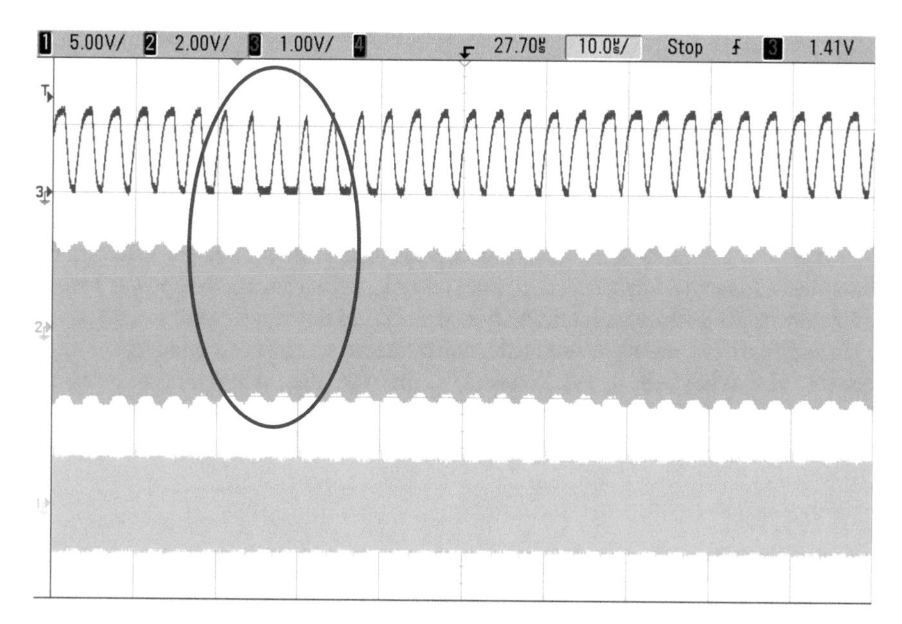

Fig. 6.12 Degradation of demodulated uplink signal (300 kbps) at the transition region from high to low when downlink data rate is increased to 10 kbps *violet* uplink demodulator output (1 V/div), *green* uplink demodulator input (2 V/div), and *yellow* rectifier input or load coil (5 V/div) (timescale: 10 µs/div) [1]

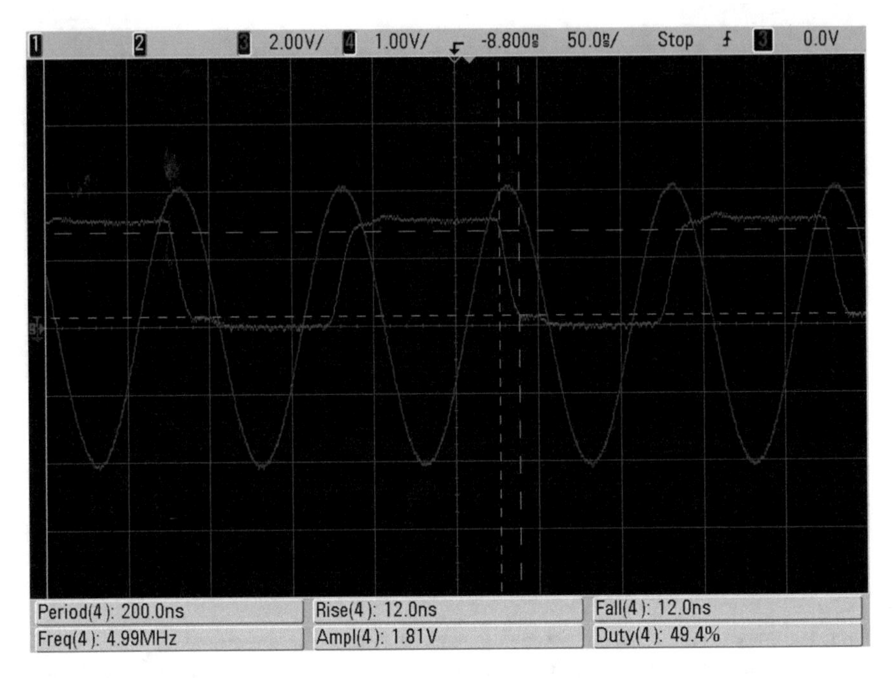

Fig. 6.13 Output of the clock recovery circuit and the input signal waveform at the input of the source coil

increased while MI is kept constant. Since increasing the modulation index reduces the power transfer efficiency, we fix it to 5% for the downlink in this situation. Figure 6.12 shows an example of the latter case in which downlink data rate is increased to 10 kbps compared to the case in Fig. 6.10. As a result, uplink data rate has to be decreased to 300 kbps if 10 kbps downlink is to be maintained. Note that degradation occurs only at the transition regions where self-referenced ASK demodulator has a lower performance. As a final remark, we have measured power transfer efficiency as 30% while maintaining both uplink and downlink communication.

In addition to wireless power transfer and bidirectional data communication, clock recovery circuit has been tested. Rise and fall times for the clock signal are measured to be 12 ns as shown in Fig. 6.13.

6.1.2 Two-Frequency Approach

This section gives the details of the performance of the system using two-frequency approach. This approach is developed to monitor multi-unit activity which requires higher data rate compared to fast ripple monitoring systems. Figure 6.14 shows the micrograph of the implant electronics of the two-frequency approach which is composed of the rectifier, LDO regulator, downlink demodulator (DeMod), clock recov-

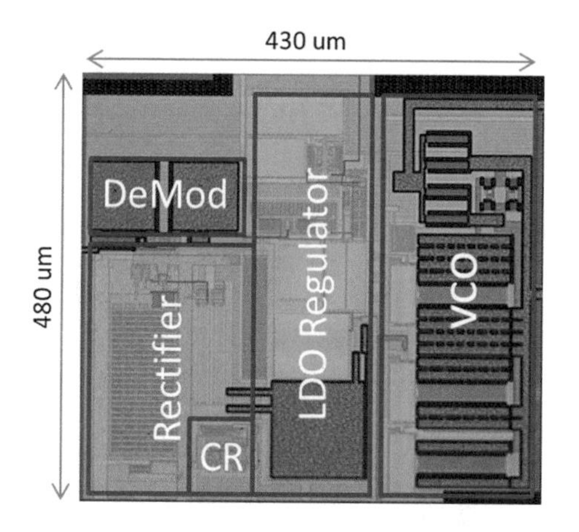

Fig. 6.14 Micrograph of the implant electronics composed of the rectifier, LDO regulator, downlink demodulator (DeMod), clock recovery circuit (CR), and voltage controlled oscillator (VCO)

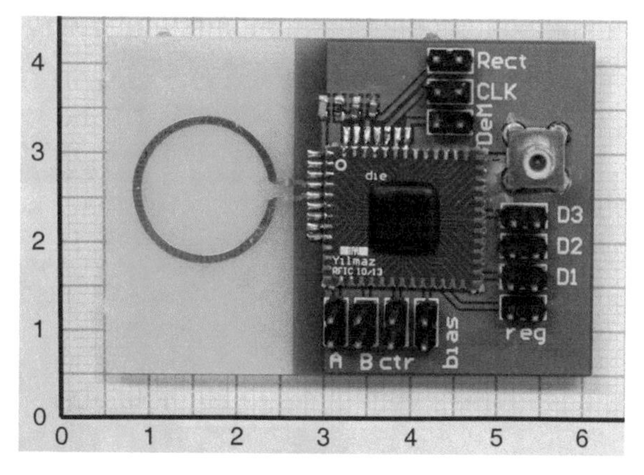

Fig. 6.15 Implant unit containing the implant electronics and the loop antenna for uplink communication (scale in cm) for characterization purpose

ery circuit (CR), and voltage controlled oscillator (VCO). In Fig. 6.15, the test PCB used for initial characterization of the circuit is presented.

Assembly of the implant part is based on stacking the following three components:

1. loop antenna of the transmitter in the implant,
2. integrated implant electronics wirebonded on a PCB, and
3. secondary and load coils of the 4-coil resonant inductive link fabricated on another PCB.

Figure 6.16 presents the front and back side of the assembled implant as well as the measurement setup used for characterization.

The dimensions of the implantable board is measured to be 20 mm × 20 mm × 1.2 mm. Note that the two PCB lines carrying the outputs to the headers are only required for measuring (input, output, and intermediate) signals. When the characterization is over, removing them does not affect the operation of the implantable part as all of the critical SMD components are placed on the implantable part. A holder

front

back

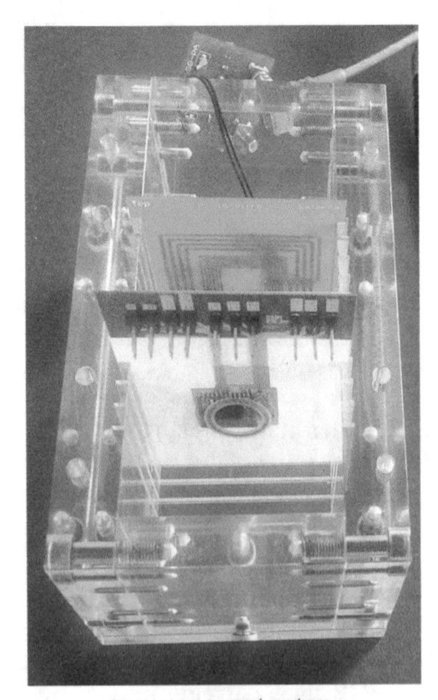

measurement setup

Fig. 6.16 Front and back side of the implant with the holder to keep it aligned with the external coils in the measurement setup

has been prepared to keep the implant unit in the corresponding slot of the measurement setup. The main purpose is to align the external coils and implant coils to maximize the power transfer efficiency during electrical characterization. Front side of the implant, which contains the secondary and load coils, faces with the external base station. Implant electronics and the loop antenna are placed on the back side of the implant.

6.1.2.1 Experiments

Characterization of the system begins with measurements of the wireless power transfer link without data communication. WPT has been realized at 36% efficiency with a separation distance of 10 mm between the external base station and implantable unit while the delivered power at the output of the regulator is 10 mW. Then, 10 kbps downlink communication has been realized with an efficiency of 33% as in single-frequency approach.

Finally, uplink communication link has been established at 430 MHz which falls into a MedRadio band (426 432 MHz). The antennas are placed with a distance of 60 cm which is close to the average distance from the head to the waist for humans, assuming that external base station can be placed onto a belt. Using on–off keying (OOK) modulation scheme, VCO has been modulated with 1.8 MHz switching rate. This rate corresponds to 1.8 Mbps data rate with Manchester encoding or 3.6 Mbps without coding. More practically, the targeted data rate (1.92 Mbps) for multi-unit activity recording is achieved. Power consumption is measured to be 830 μW, while uplink communication link is established. In addition, bit error rate (BER) has been measured to be less then 10^{-6} under the given operation conditions. BER has been measured by means of a FPGA. A pseudo-random number generator has been implemented on FPGA, and the output is fed to the OOK switch of the VCO. Next, the received signal at the output of the comparator is also fed to the FPGA in order to compare with the seed bit stream. Since there is a certain delay on the line, required amount of delay is added to the seed signal.

Figure 6.17 exhibits the waveforms acquired from different points of the near-zero IF receiver. We observe that OOK is successfully realized by observing the output of the IF output of the mixer. Logarithmic amplifier senses the power and converts it into a corresponding voltage level. Here, since the switching frequency is higher than its output driving capability, it is fortified with a high-speed comparator so that output can be fed into a digital circuitry.

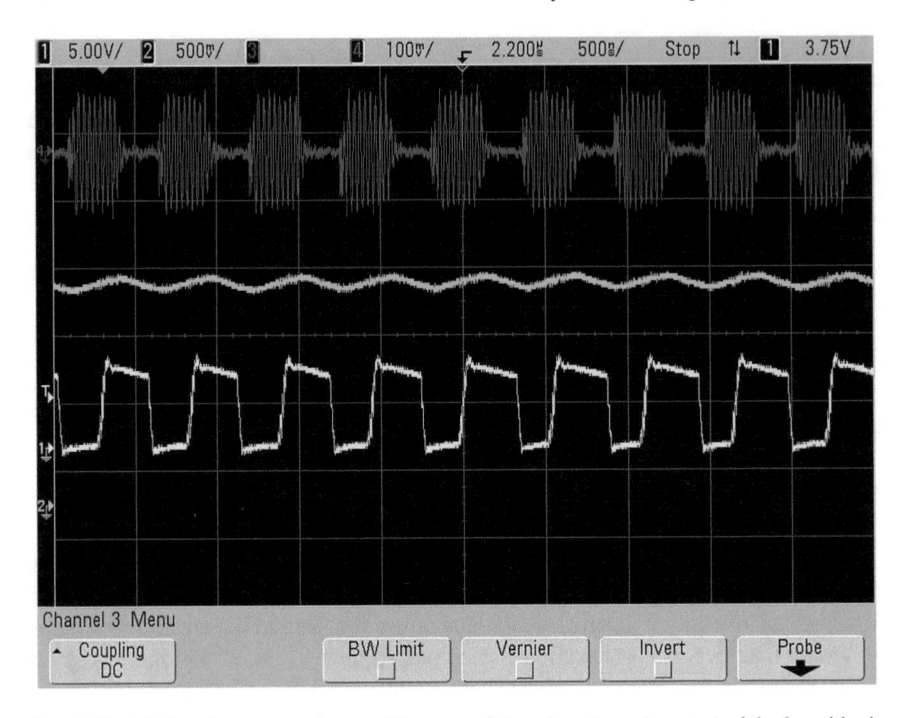

Fig. 6.17 1.8 Mbps from *top* to *bottom* IF output of the mixer (*green*), output of the logarithmic amplifier (*violet*), and the output of the comparator (*blue*)

6.2 In vitro Experiments

Following the complete electrical characterization of the wireless power transfer and data communication system, additional tests are performed in the in vitro environment. A two-chamber PMMA test setup has been manufactured as shown in Fig. 6.16 to fulfill this purpose. Moreover, the chamber containing the implant is filled with mock cerebrospinal fluid (CSF) solution to approximately model the tissue environment around the cortex. Mock CSF is composed of 125 mM NaCl, 2.5 mM KCl, 1.18 mM $MgCl_2$, and 1.26 mM $CaCl_2$ [3]. Remark that the distance is kept 10 mm; however, now the composition is as follows: 4-mm air, 1-mm PMMA, and 5-mm mock CSF solution.

Using the in vitro test setup, first the entire system is tested mimicking the operating conditions in air. Figure 6.18 depicts the power transfer efficiency of the system with respect to the data rates applied. For in vitro measurements, the power transfer efficiency does not change drastically since almost half of the medium is still air. Here, air medium is preserved for testing due to its resemblance to loose bandages which are used to cover the surgery openings.

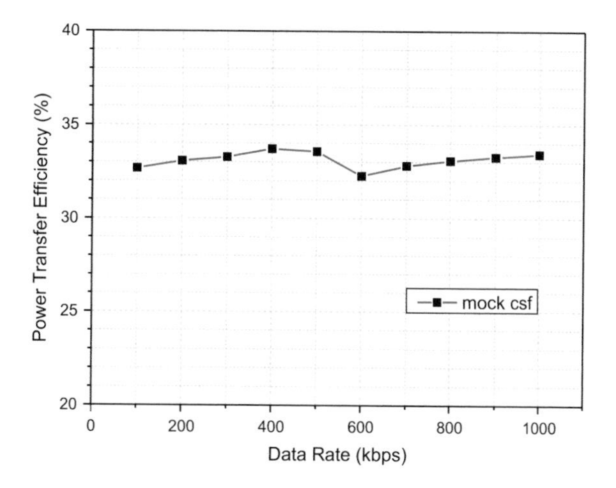

Fig. 6.18 Power transfer efficiency of the system with respect to uplink data rate in mock cerebrospinal fluid

6.2.1 Long-Term In vitro Experiments

In addition to in vitro characterization, the in vitro measurements have been conducted for 30 days to observe long-term response of the packaging. Power transfer efficiency and all other voltage waveforms are found to be constant during this period. Therefore, it can be concluded that an equivalent permeability/thickness value of 2.25×10^{-7} cm^3(STP)/s-cm^2-cmHg is sufficient to protect an implant ($14 \times 14\,mm^2$) for 30 days.

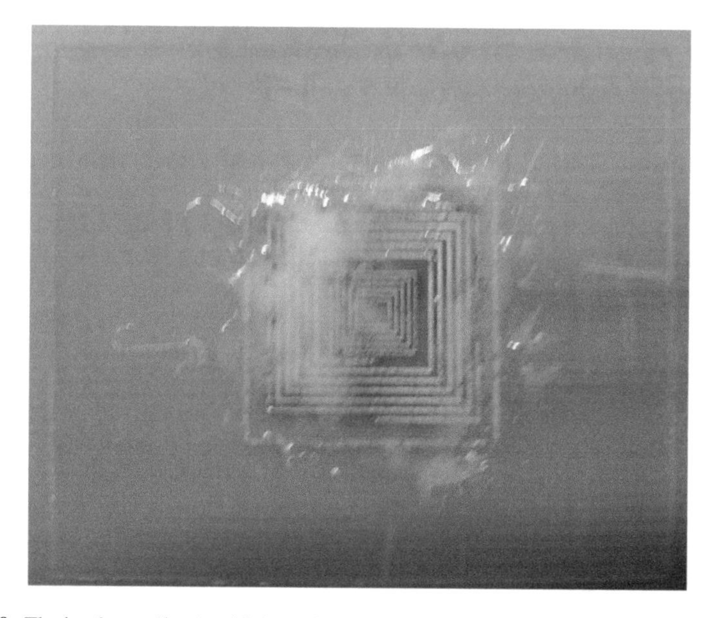

Fig. 6.19 The implant coils after 45 days of exposure to mock cerebrospinal fluid (CSF)

Following the successful operation of the implant for 30 days, the system is left working to observe further the performance of the packaging. Figure 6.19 shows the final situation of the implant coils after an exposure of 45 days to mock CSF. As observed from Fig. 6.19, a metal corrosion has occurred which causes the metal lines forming the inductive coils. Additionally, the color change in the transparent package can be attributed to the penetration of water vapor and possibly small-sized ionic compounds which penetrates the package and/or the ionic compounds as a result of chemical reactions between water molecules and metals in the implant.

6.3 In vivo Experiments

Electrical characterization of the system in air and in mock CSF solution is followed by the initial in vivo experiments on the rat brain. The experiment was performed in the Neurosurgery Department of University of Bern. Electrodes and interface circuits used in the experiment are designed by Microelectronics System Laboratory (LSM) of Ecole Polytechnique Federale de Lausanne (EPFL). Animal experiments were approved by the Animal Research Ethics Committee of the Canton Bern, Switzerland.

Intracranial EEG signals were successfully recorded from the setup presented in Fig. 6.20a–d, using the wireless-powered low-noise front-end designed by LSM [4]. During the experiments, a 16-channel floating electrode array (Alpha Omega GmbH, custom-made), consisting of sixteen recording and two additional reference and ground electrodes (each separated with 400 μm), is used.

6.3.1 Surgical Procedure for the Implantation of the Electrode Array into a Rat Brain

Two adult male Wistar rats weighing 300–400 g (Janvier, France) were deeply anaesthetized by isoflurane inhalation (75% N_2, 20% O_2, 5% isoflurane) followed by intraperitoneal injection of narketane ketaminum (75 mg/kg body weight) and xylapane (xylazinum; 5 mg/kg body weight) and analgized by subcutaneous injection of buprenorphine (0.5 mg/kg). Subsequently, the rat was mounted on a stereotaxic frame (Stoelting) and subcutaneously injected with a local pain killer (Lidocain HCL 1%, 0.2 ml). The head was shaved and disinfected before the midline incision was made. The scalp was opened and held with micro-clamps. Two burr holes were drilled at the following coordinates (in relation to the bregma): A/P $= +1$ mm and -0.4 mm, respectively, M/L $= -2$ mm, and the dura was carefully removed. The electrode reaching 2 mm into the cortex was positioned at the following coordinates (in relation to the bregma): A/P $= +\ 0.5$ mm, M/L $= -2.5$ mm (Fig. 6.20b, c) [4].

The result of this initial in vivo experiment on the rat cortex, the extracellularly recorded action potentials, is shown in Fig. 6.20d, e [4]. In these experiments, the

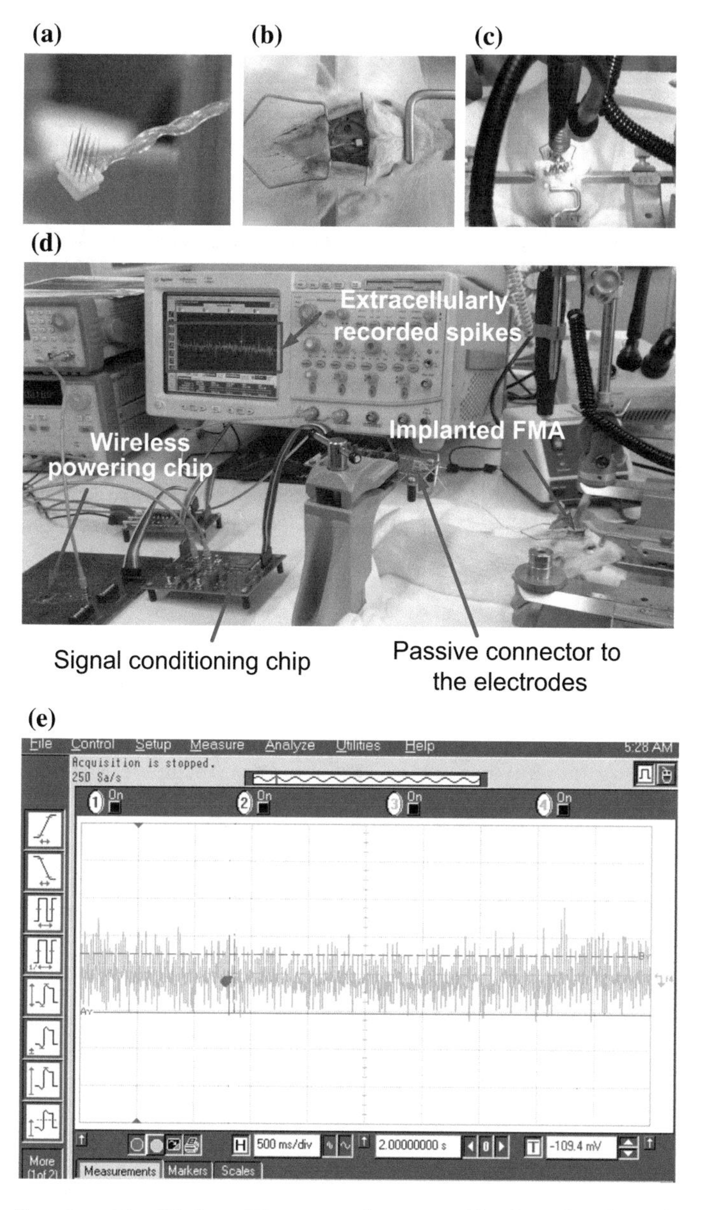

Fig. 6.20 Experimental validation of the proposed system; **a** Floating microelectrode array; **b**, **c** implanted electrodes in a Wistar rat brain; **d** measurement setup for in vivo experiments, and **e** extracellularly recorded neural signal [4]

half-wave active rectifier and the low drop-out voltage regulator, which have been implemented as a part of this work, have been validated. These circuits are employed to generate a DC supply for the low-noise amplifiers designed by LSM, EPFL.

6.4 Summary

This chapter summarizes the system integration for two system-level approaches and corresponding experimental results obtained from these implementations. Before giving the results of the complete wireless power and bidirectional data communication systems, individual performances of each block have been presented to give an insight how integrating multiple functions on the same channel affects each other and the final result. In summary, targeted specifications given in Table 2.1 have been shown to be achieved in terms of electrical characterization in air. Moreover, it is also realized in mock cerebrospinal fluid (CSF) to show that system works well with its packaging in a liquid environment mimicking the in vivo operation environment. Additionally, a long-term in vitro characterization has been realized to observe the effects of corrosion and to find out the longest time that the implant can work properly. Finally, first step of a series of in vivo experiments has been realized on a mouse brain to record neural signals while using wireless power transfer feature to supply the interface electronics. Results given in this chapter have been published in numerous articles including [1, 5].

References

1. G. Yilmaz, C. Dehollain, Single frequency wireless power transfer and full-duplex communication system for intracranial epilepsy monitoring. Microelectr. J. **45**(12), 1595–1602 (2014)
2. G. Yilmaz, C. Dehollain, Capacitive detuning optimization for wireless uplink communication in neural implants. in *5th IEEE International Workshop on Advances in Sensors and Interfaces (IWASI), 2013*, pp. 45–50, June 2013
3. T.P. Obrenovitch, A.M. Hardy, J. Urenjak, High extracellular glycine does not potentiate n-methyl-d-aspartate-evoked depolarization in vivo. Brain Res. **746**(12), 190–194 (1997)
4. M. Shoaran, G. Yilmaz, R. Periasamy, S. Seiler, and S. Di Santo et al., A low-power integrated circuit for a wireless 100-electrode neural recording system. in *Biomedical Circuits and Systems Conference, 2014. BioCAS 2014*, IEEE, October 2014
5. G. Yilmaz, O. Atasoy, C. Dehollain, Wireless energy and data transfer for in-vivo epileptic focus localization. IEEE Sens. J. **13**(11), 4172–4179 (2013)

Chapter 7
Conclusion

This book focuses on the realization of wireless power transfer and data communication as a part of a completely wireless system aiming intracranial epilepsy monitoring. The entire system aims to improve the acquired neural signal quality by replacing the analog signal-carrying wire bundles with wireless digital communication. Elimination of the wires will also improve the patient comfort and health by liberating the patient from an intensive care unit at the bedside as the current practice dictates. Additionally, it is anticipated that the risk of infection will decrease owing to the complete suture and sealing of the operated skull region post-surgery. Furthermore, the implemented autonomous wireless recording system prolongs the neural recording period. It thus overcomes the disadvantages of a system with wired recording electrodes which require bulky external medical recording equipment.

The system, composed of wireless power transfer and wireless data communication blocks, has been designed, implemented, and tested with two different system-level approaches. These two approaches require a broad interdisciplinary insight and an enthusiasm for research at the crossroads of science, engineering, and technology. Bringing together different system-level approaches with innovation at circuit level demands a diverse set of electronic engineering skills. Biocompatible packaging, implanting, and biosafety require deep insight into polymer science, mechanical engineering, and basic sciences.

Main innovations proposed in this study can be listed as follows:

- A single-frequency system, which incorporates wireless power transmission, external unit to implant communication (downlink), and implant to external unit communication (uplink), has been proposed and implemented successfully.
- A two-frequency system which uses a higher frequency and dedicated transmitter/receiver for implant to external unit communication has been proposed and implemented successfully in order to increase the uplink data rate.

G. Yılmaz and C. Dehollain, *Wireless Power Transfer and Data Communication for Neural Implants*, Analog Circuits and Signal Processing,
DOI 10.1007/978-3-319-49337-4_7

- This study is one of the first studies in the literature which uses 4-coil resonant inductive link topology for wireless implant powering. The highlights of this system-level approach are the increasing number of publications employing this topology.
- A polymeric packaging has been designed, implemented, and tested within the frame of in vitro experiments. Remarking that the system could work more than a month with this package indicates a possible use in long-term implantation.

Key points of the research that led to this book can be expressed as follows:

- Power transfer efficiency achieved under the condition that there is no data communication is almost kept at the same level when there is an active data communication on the link. This performance can be attributed to the capability of low modulation index usage.
- Power transfer efficiency has been significantly increased by designing an innovative topology originating from the world of high-power charging applications. Miniaturization of such a system poses issues such as skin effect or proximity effect which are not encountered in large-scale versions. These problems have been addressed and solved.
- Not only the short-term in vitro characterization, but also the long-term in vitro durability experiments have been conducted. Additionally, modeling of the polymer package has been done in order to give an outlook for future packaging ideas.
- First stage of a series of in vivo experiments on laboratory rats have been conducted successfully.

Index

G. Yılmaz and C. Dehollain, *Wireless Power Transfer and Data Communication for Neural Implants*, Analog Circuits and Signal Processing,
DOI 10.1007/978-3-319-49337-4

Printed in the United States
By Bookmasters